Conversations About

ASTROPHYSICS

&

COSMOLOGY

Ideas Roadshow

INTELLIGENT. INQUISITIVE. INTERNATIONAL.

Ideas Roadshow conversations present a wealth of candid insights from some of the world's leading experts, generated through a focused yet informal setting. They are explicitly designed to give non-specialists a uniquely accessible window into frontline research and scholarship that wouldn't otherwise be encountered through standard lectures and textbooks.

Over 100 Ideas Roadshow conversations have been held since our debut in 2012, covering a wide array of topics across the arts and sciences.

All conversations in this collection are also available separately. See www.ideas-on-film.com/ideasroadshow for a full listing.

Edited by Howard Burton; preface and all introductions written by Howard Burton.

Contents

A UNIVERSE OF PARTICLES:
COSMOLOGICAL REFLECTIONS
A CONVERSATION WITH ROCKY KOLB

INFLATED EXPECTATIONS:
A COSMOLOGICAL TALE
A CONVERSATION WITH PAUL STEINHARDT

THE CYCLIC UNIVERSE
A CONVERSATION WITH ROGER PENROSE

Textual Note

The contents of this book are based upon separate filmed conversations with Howard Burton and each of the five featured experts.

Justin Khoury is Professor of Physics at the University of Pennsylvania. This conversation occurred on October 22, 2014.

Scott Tremaine is Professor Emeritus of Astrophysics at the Institute for Advanced Study. This conversation occurred on October 24, 2014.

Rocky Kolb is the Arthur Holly Compton Distinguished Service Professor of Astronomy and Astrophysics at the University of Chicago. This conversation occurred on May 15, 2015.

Paul Steinhardt is the Albert Einstein Professor in Science and Director of the Center for Theoretical Science at Princeton University. This conversation occurred on May 15, 2015.

Roger Penrose is Emeritus Rouse Ball Professor of mathematics at the University of Oxford and co-recipient of the 2020 Nobel Prize in Physics. This conversation occurred on October 14, 2012..

Howard Burton is the creator and host of Ideas Roadshow and was Founding Executive Director of Perimeter Institute for Theoretical Physics.

Conversations About

ASTROPHYSICS
&
COSMOLOGY

Preface

There are many areas of scientific research that resonate strongly with the general public, from genetics to neuroscience to particle physics, but few can compare with the appeal of astrophysics and cosmology.

Perhaps it boils down to a question of accessibility. Appreciating the finer points of the structure of DNA or a proton can be daunting even for highly-trained professionals, but everyone knows what it's like to look up at the night sky in awe and wonder.

It's not that astrophysics is easy, of course—far from it—it's that, somehow, its fundamental allure is so universal as to almost not need stating. Tell someone at a party that you're studying geology and you'll likely be asked, *"So, what are you going to do with that?"* Tell people that you're studying astronomy, on the other hand, and you'll probably simply hear, *"Cool!"*

Even the most practically-minded person, it seems, can't help but feel a deep sense of personal fascination at the prospect of uncovering some secrets of the universe. And over the past few decades, things have only got more fascinating still Cosmology, once regarded as little more than speculative hand-waving, has been stunningly transformed into one of the most rigorous, data-driven fields in all of science. Dark matter has gone from being viewed as some conceptual or observational mistake that astronomers are one day going to sort out a mysterious entity that accounts for more than one quarter of the total energy of the universe. Black holes have moved from theoretically possible structures to relatively common astrophysical objects, with startlingly large ones, of masses millions or billions times that of our sun, now being thought to lie at the heart of most, if not all, galaxies.

And, perhaps most significant still, rather than the expansion rate of the universe slowing down due to gravity as most people had envisioned, a number of rigorous observations have confirmed that the universe is, in fact, accelerating in its expansion, with the associated force responsible—the so-called "dark energy"—for a shocking 70% of the energy of the universe.

Combining "dark energy" and "dark matter", then, puts us in the decidedly unexpected position where we're forced to admit that no less than 95% of the total energy of the known universe lies in stuff we really, to all intents and purposes, are entirely ignorant of.

Talk about awe and wonder.

I'd like to be able to assure you that the answers to all of these questions are contained within this collection of conversations with scientific experts, but of course I can't.

What I *can* tell you however, is that each of the five conversations offers a remarkably unique opportunity to get a vivid, first-hand perspective of what it's like to be at the very forefront of this riveting, rapidly-evolving scientific enterprise.

Justin Khoury reveals the often-considerable gap between formal scientific positions and personal scientific desires.

"There's a funny discrepancy between what people talk about at the conferences and the formal talks. We give talks that are rather conservative, because that's how we're trained. That's a good trait to have, actually, to be conservative: we're not completely wild.

But at the end of conferences, sometimes somebody will want to have some fun and say, 'Let's have a survey, let's write down some options for what we think dark energy is'.

"And it's always startling that the way people vote on these surveys

is far from conservative: they vote for the most radical option. Now, why is this? Well, maybe it's because we are hopeful: we don't want to put our names on a paper that's crazy, but deep in our heart, we hope that it's a radical answer so we have more stuff to do for the next 20 years. Or it could be a real belief: we don't have an idea for how to actually explain it and write papers about it, but we actually believe that it's true."

Scott Tremaine illustrates how the process of discovery in astrophysics is typically very different from other domains of physics.

"In a certain sense, astrophysics is more like a detective story than other branches of physics, because in other branches of physics if you have something you don't understand, you try to design an experiment that's going to allow you to understand it. In astrophysics, however, you often can't design any experiments. You have a much more incomplete set of clues; and, like Sherlock Holmes, you're trying to deduce what must have happened from partial evidence. Also, like in Sherlock Holmes, the game of knowing when you've got the answer—because one answer is so much more compelling or simple or beautiful than all the other possibilities—is often more sophisticated in astrophysics than in other branches of physics.

You have to have the imagination to ask if there are other things that are allowed by the laws of physics that we haven't detected which might be quite different; and, if so, should we have seen them already, and are there techniques for detecting them?"

Rocky Kolb admits how, dark energy, in stark contrast to dark matter, simply drives him nuts.

"Dark matter doesn't really bother me. I don't lose sleep over it. My attitude is, 'Oh, this is a great opportunity to use this idea. Maybe it's this, maybe it's that.' Dark energy, on the other hand, to me is like fingernails on the chalkboard. It just drives me nuts. I don't like it. I don't like it; I admit that it's a prejudice, but there it is.

I don't have a good explanation. It's not a logical thing. It's not that I say the observations are wrong—although I did for at least a couple of years. I kept saying, 'There must be some other effect responsible for these observations', until I was finally convinced otherwise. I just can't swallow it."

Paul Steinhardt describes his profound bemusement at how many of his colleagues flatly deny the multiverse problem of inflationary cosmology.

"I've often had this discussion where I'll say, 'Well, what do you think about the multiverse problem?' and they reply, 'I don't think about it'.

*"So I'll say, 'Well, how can you **not** think about it? You're doing all these calculations and you're saying there's some prediction of an inflationary model, but your model produces a multiverse, and so it doesn't, in fact, produce the prediction you said: it actually produces that one, together with an infinite number of other possibilities, and you can't tell me which one's more probable'.*

"And they'll just reply, 'Well, I don't like to think about the multiverse. I don't believe it's true'.

*"So I'll say, 'Well, what do you **mean**, exactly? **Which** part of it don't you believe is true? Because the inputs, the calculations you're using—those of general relativity, quantum mechanics and quantum field theory—are the very same things you're using to get the part of the story you wanted, so you're going to have to explain to me how, suddenly, other implications of that very same physics can be excluded. Are you changing general relativity? No. Are you changing quantum mechanics? No. Are you changing quantum field theory? No. So why do you have a right to say that you'd just exclude it?'"*

And **Roger Penrose** relates his longstanding frustration with the fact that so many of his colleagues fail to appreciate the importance of why the universe started out in such a remarkably smooth state, maintaining that the answer usually proposed to account for it—cosmic inflation—actually does no such thing.

> *"The original theorem that I had on black holes showed that singularities came about no matter how irregular the collapse was, or whatever kind of matter you had, just so long as the energy densities weren't negative. And then Stephen Hawking picked up on this and applied these techniques to cosmological situations and we collaborated on a paper that encompassed most of these results.*

> *"But there is a sort of irony there too, because at the time I was thinking, 'Why are we limiting ourselves to these simple models of the universe? We could have all sorts of complicated things. That's why you need the singularity theorems.' But the point is that the universe is not like that.*

> *"I remember being in a car near Princeton going to some conference, and in this car was Jim Peebles, one of the world's most distinguished cosmologists. And I was saying, 'Surely people should have considered all these complicated things that might happen...' And he replied, 'But it's not like that. The universe is this very uniform state.'*

> *"That is what started me thinking that this was the **real** puzzle: why is the universe so smooth and uniform instead of such a great big mess?*

> *"Almost all of the calculations that people do are in a background of a very smoothed out universe. They put in a little perturbation here or there, but that doesn't really come to terms with the problem. If we had been in this unbelievable messy situation to begin with then inflation wouldn't do anything for us at all. The 'unbelievable mess' would have been a state of enormously high entropy—in terms of the gravitational degrees of freedom—and inflation, being a time-reversible dynamical process acting in accordance with the Second Law of Thermodynamics, wouldn't be any use at all: it would just spread out the clumps. So it's really no explanation to the question of why our universe is so uniform."*

Fully appreciating all the subtleties associated with today's deep cosmological mysteries is unquestionably difficult. But getting a genuine taste of what the world's top cosmologists are grappling with turns out to be startlingly easy.

Cosmological Conundrums

A conversation with Justin Khoury

Introduction

Into The Light

Surprising though it might seem, most physicists have strong revolutionary sympathies. From Galileo to Kepler, Newton to Einstein, the most successful natural scientists in history have all tended to buck the established wisdom of their day as they boldly led us towards profoundly deeper levels of understanding about the world around us.

The most recent revolutionary period occurred a little more than a century ago and was perhaps the most profound of all, consisting of two near-simultaneous overthrows of conventional wisdom that eventually resulted in today's twin pillars of relativity theory and quantum mechanics, completely overhauling how we look at space, time, matter and energy.

A characteristic feature of revolutions, however, is that, most of the time, you don't see them coming. Lord Kelvin, an impressively accomplished scientist who did fundamental work in both electricity and thermodynamics (and after whom the Kelvin temperature scale is named), is now perhaps best remembered for one of the most ill-timed prognostications in scientific history.

"*There is nothing new to be discovered in physics now,*" he breezily declared to a group of physicists at the British Association for the Advancement of Science, "*all that remains is more precise measurements.*

That's a dangerous sentiment to voice at any moment, but Kelvin's timing turned out to be particularly disastrous, given that he gave this speech in early 1900, only 5 years before Einstein's *annus mirabilis*, and mere months before Planck's blackbody radiation postulate that led inexorably to quantum theory.

Such lessons from history are uppermost in the mind of University of Pennsylvania physicist Justin Khoury these days. Justin works at the interface of particle physics and cosmology; and, as such, finds himself enmeshed in two of the greatest scientific mysteries of our, or any, age: dark matter and dark energy.

> *"We don't know what these entities are, but there's a proposal out there that dark matter consists of weakly interacting particles and dark energy is a form of vacuum energy; and those ideas, by-and-large, work well against the data.*

> *"Now there's a sense—a fairly widespread sense—that the field is in this butterfly-collecting mode: that we just need to cross the t's and dot the i's and we're reaching the end. But, of course, that's a pitfall, as we know from history; one has to be careful.*

> *"It may be that this is what dark matter and dark energy is, but it could also be that we're in for some surprises."*

But often, in order to become confronted with the unexpected, one has to be prepared to look for it in the first place. And one thing that particularly concerns Justin is that, as cosmology has steadily moved from speculative musings into a rigorous data-driven domain on the back of decades of increasingly precise observations of the cosmic microwave background, it has fallen prey to an increased unwillingness to countenance views outside of its established mainstream.

> *"One of the downsides to having all of this data and the science maturing, if you will, is that the field has also become more conservative. Maybe that's a good thing, because we want to converge at some level, but there is a definite worry now that it's become less accepting of new ideas. New ideas are never nicely packaged when they come out—they're typically ill-formed. And if they're constantly being shut down, then we may be missing out on something.*

"That's one thing that I've definitely observed from the time I was a graduate student. Back then, we came up with this crazy idea about the early universe, and, at the time it was more or less—I wouldn't say accepted—but there was a certain open-mindedness about us suggesting it. Whereas I think that nowadays, 15 years later, such an idea would be much harder to propose.

"Maybe it's a natural pendulum swing: before it was the Wild West and now we're in the conservative part of the swing, but I hope there will be a 'market correction', where we go back to a more open-minded attitude. It's important, because the new ideas will come from young people.

"That's usually the case, that young people will transform the field. I don't want to be in a situation where, a hundred years from now, we discover what the correct theory was and we realize that we ignored this young person today who had the right idea—we shut her down. There are examples like that throughout history, and I just hope our field won't be like that."

Justin's concerns aren't based simply on principles of fairness or a well-honed historical awareness of the dangers of scientific hubris. He has a hunch that another profound scientific revolution is just around the corner, and all that is really needed to push things over the edge is a large infusion of young, dedicated, iconoclastic scientists.

"I think that we're about to experience the same sort of revolution that happened at the beginning of the 20th century with the invention of relativity and quantum mechanics.

*"If you're a young person, I think it's the perfect time to jump into this field and contribute something original. The problem with someone who's been in the field for a long time like myself is that we get blasé—we think, **"Oh no, this idea will go away"** and that sort of thing, but young people don't have that, and I think that's really helpful and refreshing.*

"So my message to young people out there is: Don't be afraid to come in, propose new ideas, think outside the box and think

about alternative systems that we've thought about in history that connect these different phenomena. We need fresh blood."

Aux armes, citoyens! You have nothing to lose but your preconceptions!

The Conversation

I. Becoming a Physicist

The power of passion

HB: Let's start with how you got interested in physics, and cosmology, in particular. How did that all begin for you?

JK: Well, I come from a rather modest background: we were not an academic family. I grew up with my mom and my grandmother, and, although they were not academics, they were always—my mother still is—incredibly curious; and I think that's the most important aspect of my childhood.

We would regularly buy new encyclopedias and new, illustrated dictionaries. My grandmother was incredibly curious, and I fed off that. In my family, there was a sense of pride that schooling was important, and doing well at school was important. Of course, like every Canadian boy, my goal was to become a hockey player, but I quickly realized that I couldn't skate, so I had to look for something else.

I had different passions in high school. I loved music—in fact, I went to a music school. I loved the piano, but I also loved science. Ultimately, it was a practical thing: practicing the piano for an hour was incredibly painful, even though I loved it, while doing physics and math homework was sort of fun, even for hours on end.

Another interesting aspect for me was related to the power of communication and outreach—like what you're doing with *Ideas Roadshow*.

I remember, as a child, being highly influenced by a documentary on Einstein's life: his discovery of relativity and the impact it had on society. It had this huge impact, and the romantic view of trying to seek the fundamental laws of the universe has stayed with me ever since.

Nowadays I teach general relativity. Tomorrow I'll teach Einstein's theory in my classroom; and after our conversation, I'll go to a café and think about these ideas in my own work. I actually get paid to do that, which I think makes me incredibly fortunate. Academics like to complain like everybody else, but once in a while, we have to step back and realize how great our jobs really are.

HB: You spoke about the importance of curiosity, and mentioned your wide interests in addition to science. Why physics and mathematics in particular? Was it just this documentary about Einstein or was it some sort of specific orientation or ability?

JK: It was a few random events for sure. One was definitely that documentary, but another was the power of high school teachers. You can talk to anybody in the field, and you'll invariably discover that he or she had at least one very influential high school teacher.

In high school, I was, in all modesty, relatively good in French. I was better in French, I'd have to say, than I was in math and physics, but I had one teacher who pushed me and made me realize that I could do math better than I could do French literature.

The power of teachers is incredible: that really channelled me towards physics. Now, why physics as opposed to biology or some other field of the natural sciences? I don't know. I guess what I always liked about physics was that it went to the deepest, fundamental level of nature, and it was also incredibly quantitative.

That's one thing I've always liked about it: if I do an exam and I get the right answer, they can't take any points off, but if it's an English essay, say, it's in the hands of a subjective professor.

So for me, that was always something appealing—that my grades were in my own hands—which is a very practical thing, but at some level it's true.

HB: So, you had this formative influence from your high school physics teacher who both encouraged you to plunge into mathematics and physics at a higher-level and also discouraged you from devoting your life to French literature—

JK: Yes; and I'm forever grateful.

HB: Right. And then you went to McGill for your undergraduate degree?

JK: Yes. I grew up in Sherbrooke, but I went to college at McGill University. That was pretty hard, actually. I was never the child prodigy that got into college and scored by far the best grades with ease. I wouldn't have won the Olympiads for example, not at all.

But I tried physics and thought that, if it didn't work, I could always do something else. But I really loved it. I had a passion for it.

When young people think about academia, or, say, having a professional life in physics or in math or what have you, there's a tendency to think that it's all about talent: the person who scores the highest in the exams early on will be the one who succeeds.

But it's not about "pure talent" per se; it's a marathon, not a sprint. It takes 10 or 15 years from that point on until you're a settled professor—it's incredibly difficult, and most people will give up.

Looking back at it, it's really a question of determination as much as anything else. I mean, you have to be good at calculus, but then again, a whole lot of people are good at calculus, so you just have to keep on plowing ahead—and that's all passion. But McGill was difficult; it was a lot of hard work.

Then I got to Princeton for graduate school; and then, somehow, things clicked. I think the transition is from the modus operandi of saying, "*I have to solve problems that are given to me*," to that of having to find your own problems, which is what research is all about. And I think when I made that transition, that's really when I felt the most comfortable.

HB: This makes me think of a conversation I had with David Politzer at Caltech recently. He was talking to me about his experiences with Richard Feynman. Of course, there's this mythology associated with Feynman: that he could do everything, solve every problem and was so incredibly broad.

He *was*, of course, remarkably broad in his interests and clearly unbelievably talented. But what David told me was that, when he or someone else asked Feynman a problem, Feynman would invariably go back and look at his notebooks, where, more often than not, he'd find a solution that he had already calculated long ago—because the guy was calculating things all the time.

And that's the key point: he was so passionate about so many things, and so curious, and so determined to try to work out everything all the time, that very often he had already solved something, or at least given it a lot of thought.

As you say, there's this mythology out there that there are just a few geniuses who just have "the gift", but that doesn't match up with my experience. And if it doesn't even apply to Feynman, it's hard to know who it would apply to.

JK: That's right. There are these "rules" that people talk about now, a minimum number of hours that leading sportsmen or leading academics or what have you have to put in to "make it". They all have to work very hard.

Questions for Discussion:

1. How might we encourage more high school students to "follow their passions"? Are there any actions that can be undertaken structurally, or does it all simply come down to individual teachers and students?

2. Might there be a way to use modern communications technology to scale the impact of inspirational multimedia, such as the Einstein documentary Justin mentions in this chapter that had such a strong effect on him?

II. The Victim of Its Success?

Swinging scientific pendulums

HB: When I was an undergraduate in the 1980s, cosmology had a bit of a stigma attached to it. There was this sense that it was more like philosophy, that it shouldn't be taken seriously because it wasn't a real science. People had worked out various different models of different possibilities, but the general feeling was that, as a physicist, you should really be devoting your time and attention towards topics where you could either develop concrete mathematical structure or had ready empirical overlaps so that you could go ahead and perform some real calculations.

Cosmology was looked upon as this quaint area where most of the fundamental work had been done and that's all there was to it. There was a general sense that, maybe at some point in the very far, distant future we might have some answers as to whether the universe was actually open, closed or flat, but that was about it.

And one of the most striking things for me, sociologically-speaking, is seeing that view change so drastically, to the extent that cosmology has not only become an active and highly respected field of physics, but clearly one of *the* most active and dynamic areas in all of science: the transformation in our level of understanding now of the early universe compared to thirty years ago is nothing short of overwhelming.

Meanwhile for you, it must have been just a fantastic time to be entering the field. As most successful people, one of the most important things is good timing; and you certainly seem to have had it.

JK: That's right; and I completely agree with you. The field has changed considerably even in just the 15 years that I've been in it, becoming a very quantitative science with many observations, which is incredibly exciting. These wild ideas that people had 30 years ago are now being tested and ruled out by the data, so it's an incredibly gratifying endeavour.

At the same time, though, there's a price to pay for that, which I do sometimes think about. Although I wasn't there 30 years ago, the field was more like the Wild West: people were coming up with all these crazy ideas, and there was a relatively broad sense of acceptance of, if not the specific ideas themselves, at least the motivation to propose them in the first place.

One of the downsides to having all of this data and the science maturing, if you will, is that the field has also become more conservative. Maybe that's a good thing, because we want to converge at some level, but there is a definite worry now that it's become less accepting of new ideas. New ideas are never nicely packaged: when they come out, they're typically ill-formed. And if they're constantly being shut down, then we may be missing out on something.

So that's one thing that I've definitely observed from the time I was a graduate student. Back then, we came up with this crazy idea about the early universe, and at the time this was more or less—I wouldn't say accepted—but there was a certain open-mindedness about us suggesting it. Whereas I think that nowadays, 15 years later, such an idea would be much harder to propose.

HB: It seems to me that there are at least two separate effects going on. One is that there are certainly more data, which means there are all sorts of ways by which we can rule theories out, and rule them out at an earlier time.

But the other, which I think you're touching on, is that such a situation in itself leads to a different sort of mental conditioning, insofar as people are narrowing their focus to say, "*We shouldn't really be engaging in these wild, speculative quests, we should only be thinking about things that explore this particular set of data or these particular ideas."*

And in this way, things are moving towards a more conservative view in terms of an overarching outlook or approach.

JK: Yes. So let's take the questions of "dark matter" and "dark energy". We don't know what these entities are, but there's a proposal out there that dark matter consists of weakly interacting particles and dark energy is a form of vacuum energy; and those ideas, by and-large, work well against the data.

So, now there's a sense—a fairly widespread sense—that the field is in this butterfly-collecting mode: that we just need to cross the t's and dot the i's and we're reaching the end. But, of course, that's a pitfall, as we know from history; one has to be careful.

It may be that this is what dark matter and dark energy is, but it could also be that we're in for some surprises.

I think that the conservatism is both driven by data, as you were saying, and also from the fact that the theoretical ideas are more refined, that people have higher standards of aesthetics for the theories we write down.

That's certainly a good thing, but I also think that, from my vantage point at least, the pendulum has swung a bit too far in that direction. I think we should be more open-minded, given that we don't know, in truth, what these phenomena actually are.

Questions for Discussion:

1. What do you think Justin means, exactly, when he says "we want to converge at some level"?

2. To what extent do you think the availability of data determines the sort of people who go into a scientific field to start with? Might the increased conservatism Justin mentions be more of a reflection of the dispositions of those who are currently attracted to the field?

III. Periodically Fiery

Calculating colliding branes

HB: So, let's talk a little bit more concretely about some of those ideas you just mentioned that you were involved with as a graduate student, and then I'd like to return to dark matter, dark energy and other issues.

Another more general aspect that I'd like to accomplish by this conversation is to give people a window into the world of what a practicing, theoretical scientist, theoretical cosmologist, is doing all day long.

By way of illustrating that, let's start at the beginning of your research career, your PhD and the work that you just alluded to: these "crazy"—or perhaps not—ekpyrotic ideas.

JK: Sure. When I was a PhD student, I started out in string theory. At the time, in the late '90s, that was the hottest thing in theoretical physics, but personally I had a growing sense that, though it was very interesting, it wasn't directly tied to data, which left me a little bit unsatisfied.

Then, serendipitously, I came across my eventual adviser Paul Steinhardt, a wonderful man, who was working at the interface of particle physics and cosmology. At the time, there was a sense that string theory was maturing as a science, and it could have something very important to say about the universe, about cosmology. That was what motivated us. Paul had been working on this idea that maybe the universe, or the Big Bang as we know it, arose from a collision of so-called "branes" in some extra dimension, where these surfaces—these branes— would slam into each other.

So, that was a very exciting quest. The idea was definitely provocative. Ultimately, the initial theory that we had in mind was perhaps not as successful in the way we had anticipated, but it was certainly a very fun endeavour.

In general the motivation was that, if you look back in the '60s and '70s, there were huge developments in particle physics which fed into cosmology not long thereafter. And we had this hope that maybe something like that would happen with string theory, which gave rise to this notion of the *Ekpyrotic Universe*.

HB: Where did that name come from, exactly?

JK: Well, we had this theory in place of these colliding branes, this fiery event, and we were looking for a name for it. As you know, there's a lot of marketing going on in theoretical physics: if you don't have a good name, you're not going to get citations for your paper.

So Paul, being acutely aware of this, actually went to classics professors in Princeton and described the basic scenario to them. Remarkably, they thought that it reminded them of ekpyrosis, which was a cosmological model of the Stoics, in which the universe would arise out of a great fire—and, in fact, this fire would occur periodically. So, we thought that was a good name. It reminded others of some sort of skin disease, but we liked it.

HB: So, you have this theory, and you have this possibly advantageous name associated with it, but for someone who hasn't studied these things, there are obvious questions like: "*Well, how do you know if this theory is correct?*" and "*What are you actually doing when you're writing these papers?*" Just saying that there are these things called "branes" that smash together to somehow produce a universe is one thing, but there's a lot more to it than that.

JK: Yes. The biggest obstacle to making this work was to get a spectrum of fluctuations of these branes as they collide. If they were exactly flat and parallel, of course, the collision would be simultaneous and you would get the same temperature everywhere in the universe as the universe reheated. But, in fact, as the branes come together, they will have slight ripples, because of quantum mechanics, and the collision would happen at different times in different places in the universe, which today would appear to us as different regions having slightly different temperatures, which is what we observe in the cosmic microwave background (CMB).

The biggest challenge was to calculate what these fluctuations looked like and whether they would actually match what we observed in the CMB. That's what I spent days and days doing, trying to work out the calculations.

In retrospect, it was a really risky project. In fact, I know Paul had qualms about giving this project to a student, because he knew that a student would be devoting 100% of his time for a year or more, and if it didn't work—which could well have happened— my thesis would have been considerably delayed.

So, I took a risk in some sense, but I thought that it would be a good reward if it worked out. In the end, I remember the afternoon when I did the computation and it all worked out, or so it seemed at the time, anyway: we got a spectrum that had the desired properties, which was very exciting. I still have the notebook with the equations boxed a few times, and I vividly remember going home that day knowing that the calculation had worked.

Questions for Discussion:

1. Why do you think so many physicists are convinced that there is a connection between theories of the very small (particle physics) and those of the very large (cosmology)? What, if anything, does this imply about the typical world-view of physicists?

2. Should professors be routinely trying to guide their PhD students away from "risky" projects?

IV. The CMB

Almost completely homogeneous

HB: OK, I'm naturally keen on talking in more detail about your research ideas, but before we continue, let's discuss more detail about the cosmic microwave background (CMB), starting way back at the beginning with a very brief description of what this thing is and what, specifically, we've learned about it.

JK: Well, the microwave background is the most important piece of information that we have that really led to this dramatic growth of the field in the last 20 or 30 years.

What we have at our disposal is basically, as the universe was expanding and cooling down, it eventually became cool enough to form neutral hydrogen. Protons and electrons combined to form neutral hydrogen, and at that point the universe became essentially neutral, which meant that light—photons—could propagate over large distances unimpeded.

So, we have this beautiful snapshot of the universe with this distant, relic light from when the universe was 400,000 years old. Of course, now it's 14 billion years old, so it's the equivalent of an elderly person having a picture of herself as a fetus: it's an incredible amount of information.

One thing we learned from the microwave background is that it's incredibly isotropic, which means that, if you look in all directions of the sky, the temperature is the same to one part in ten to the five: it's remarkably constant throughout all parts of the sky, which tells us that the universe back then was incredibly homogeneous.

But these very tiny fluctuations in temperature in different places of the sky have particular correlations among them: they're not completely random.

Our theories cannot predict the exact pattern that we see but, statistically, they can say certain things about that. One of the things that we observe is that these tiny fluctuations in the temperature are correlated in a way that is independent of scale.

That means that, if you were to take a picture of the microwave background and zoom in and keep zooming in, you would see, statistically, the same pattern. That's a remarkable feature of the universe; that is a very hard feature to understand from an underlying theory. Where we think these temperature inhomogeneities came from is basically the Big Bang, or near the Big Bang, or, in the case of the theory we were proposing, they would have been the result of an event before the Big Bang, this brane collision.

The big challenge in these calculations for anyone who wants to come up with a theory of the early universe is, *"How did these initial temperature perturbations come about, and why do they have this feature that they're invariant under scale?"* That's the hardest thing to come up with.

When I was working on these brane collisions I was talking about earlier, that was the goal, that's what I was trying to accomplish. And the speculation is that the interaction with the branes approaching each other all took place before the Big Bang.

Now, normally you would say that you cannot describe the universe before the Big Bang, the Big Bang is the place where the laws of physics as we know them should cease to hold, so we had to make some speculations. But based on some educated guesses as to what would happen, we could at least say how a spectrum of brane fluctuations would translate into a temperature-fluctuation spectrum in the microwave background.

Questions for Discussion:

1. What, exactly, do you think Justin means when he says that the scale invariant nature of the temperature fluctuations in the CMB is "very hard to understand from an underlying theory"?

2. What sort of "speculations" could you imagine making about the laws of physics before the Big Bang? What sort of conditions or constraints might be applicable to such laws under those very odd circumstances?

V. The Process of Discovery

Typically messier than you think

HB: And you mentioned earlier that you thought there was a much greater tolerance in the physics community to support these sorts of speculations you were indulging in back then there would be today.

JK: Yes. I've had discussions with some of my colleagues, and there's definitely this sense that the community as a whole is becoming more conservative. I think that's a worry. We all know how academia works: it's a hierarchy, it's a pyramid. You start as a young student or postdoc, fighting for your life to get the next job, hoping eventually to make it up to a professorship. How do you do that? Well, you have to impress the older folks. How do you impress the older folks? If they're open-minded, you can impress them with new ideas. As it happens, some of my older colleagues are very open-minded, but on average, right now, if you're a younger person, it's harder to propose new ideas than it used to be.

We have to be humble. What we do—although it's connected to data, as we've said—is still highly speculative. We don't know for sure, so we have to have this sense of humility.

Do we really know that dark matter exists? Do we know that dark energy really is vacuum energy? Do we know that the early universe came out of something that happened shortly after the Big Bang? Cosmology is a fossil science: we're trying to make up the story that led to what we see today.

So, like I said, maybe it's a natural pendulum swing: before it was the Wild West and now we're in the conservative part of the swing, but I hope there will be a "market correction", where we go back to a more open-minded attitude. It's important, because the new ideas will come from young people.

That's usually the case, that young people will transform the field. I don't want to be in a situation where, a hundred years from now, we discover what the correct theory was and we realize that we ignored this young person today who had that very idea—we shut her down. There are examples like that throughout history, and I just hope our field won't be like that.

HB: A point worth emphasizing, I think, that you made earlier is that, at the forefront of discovery, things are messy. But by the time it filters down to students they get presented with this beautifully ordered "package" of the way we understand nature to be.

You take a course in electromagnetism, say, when you're in first or second year and you get presented with Maxwell's equations and this beautiful structure of the way this particular set of phenomena works and the way everything naturally fits together.

But of course that wasn't actually the way it developed; that's what you get by the time it's cleaned up, analyzed and assembled in this very nice, compelling beautiful and well-integrated picture.

So that can be inspiring if you're a student, because you're presented with all this beauty and structure, but it can also be somewhat depressing because you naturally think to yourself, *"Gosh, I would never have been able to come up with this wonderful, perfect framework."* But, of course, nobody did; that's not the way it actually happened.

JK: That's right; it's deceiving. We're now approaching the 100th anniversary of general relativity, and that's another perfect example. Einstein struggled for a decade trying to come up with general relativity, and along the way he had all kinds of wrong ideas—he convinced himself of ideas that turned out to be wrong. It was hardly nicely pre-packaged, as you say. In the end, of course, it was a beautiful output that he came up with, but the process is very messy.

Questions for Discussion:

1. Should the history of science be more emphasized in the teaching of science?

2. Why do you think that important new ideas in the sciences typically come from younger people? Is the situation different in the humanities? If so, why might that be?

VI. Learning from History

Missing mass or missing theory?

HB: Let's move now to dark matter and dark energy.

We've already thrown those terms around a lot; and, I think, in the public consciousness, there's sometimes some confusion, because they both—thanks to you guys—start with "dark," which is not, after all, terribly descriptive.

So let's separate them and talk a little bit, in turn, of what each of them means, how they came to be and what our present understanding is.

Then we can talk about some possibilities, some solutions, what you've been working on, what your gut feeling is and so forth. Let's focus on dark matter for the moment.

JK: OK. The first thing to say is that we are an underrepresented minority in the universe. The stuff that we're made of—the protons, neutrons, electrons and so forth—only makes up 5% of all of the energy and matter that's in the universe. And we think that dark matter makes up 25%.

HB: This is such a shocking claim that I really think it needs to be underscored.

Again, going back to the antediluvian days when I was an undergraduate, if you were to say that *all* of our theories of physics and science more generally, everything from the standard model of particle theory to chemistry, biology, and all the rest—everything that we know about, in short—would only pertain to 5% of the stuff in the universe, you would have been considered completely crazy.

JK: Yes, it's a humbling fact. But for the remaining 95%, we're not completely in the dark—no pun intended. For dark matter, we have good confidence that we know what it is, with some caveats that we'll get into later.

In many ways, dark matter behaves like the ordinary matter that we're made of: it collapses under gravity, it clumps, it forms structures that, later on, can be hosts for galaxies. The only apparent difference seems to be that it doesn't emit light, which is why we call it dark. But in many ways, it is similar to ordinary matter.

Dark energy, on the other hand, is completely different—and that, I think, we understand even less about.

HB: We'll get to dark energy in a moment, but first of all, here's a fairly common question that almost every non-specialist will ask: *How do we know it's out there? Why are you so convinced it exists at all if we can't actually see it?*

JK: That's a key point. So far, we only observe the effects of dark matter through gravity. We're doing experiments to try to detect it otherwise, but for now we can only observe it through its gravitational effect.

That's a very humbling thing to think about, because it means that we have this phenomenon which disagrees with what we'd expect to happen, given our theory of gravity, so we infer that there is "missing matter".

It means that we have to have incredible faith in the theory of gravity, because that's what leads us to postulate the existence of this missing matter to begin with.

Einstein's theory of gravity is beautiful, we've tested it very well in the solar system and some other systems as well, but at the end of the day, it's an extrapolation: we're extrapolating a theory that works well in the solar system to galactic scales and even larger scales.

The other option would be to say, "*Well, maybe, in fact, there is no dark matter and maybe the theory of gravity itself must be revised."* There are examples of both of these sorts of approaches in the history of science.

Perhaps the most famous example involved the 19th-century French astronomer Urbain Le Verrier. At the time, there were anomalous observations of the orbital motion of Uranus that deviated slightly from what Newtonian gravity predicted. And that led Le Verrier (as well as John Couch Adams, in England) to postulate the existence of another planet beyond Uranus, which they called Neptune.

And, indeed, shortly thereafter, they did discover Neptune. So, this was a case of "missing mass", and we found it in the planet Neptune. It explained this gravitational anomaly that we were struggling with, and Newtonian mechanics was perfectly fine.

The counterexample—which again, it turns out, involves Le Verrier—is that he then discovered that the orbit of Mercury also had an anomalous aspect to it: the elliptic orbit that it traced out as it moved around the sun was precessing, very slowly, in time at a slightly different rate than was predicted by that of Newtonian gravity.

This led Le Verrier and other astronomers to postulate the existence of another planet closer to the sun, which they called Vulcan. After all, it worked before, so why not apply the same approach?

But that turned out to be wrong. Now we know the answer is Einstein's theory of general relativity, that's what needs to be used to account for the exact, observationally detected rate of the precession of Mercury's orbit.

In this instance, then, we needed to revise our theory of gravity, go beyond Newton's formalism, in order to solve the problem.

So, the question is: which one of these two situations are we in today with dark matter?

Questions for Discussion:

1. Why does Justin call the notion that our current laws of gravity work on large scales "an extrapolation"? Can you give other examples in contemporary physics where a "scale factor" exists?

2. What do you think Justin means, precisely, when he says that "Einstein's theory of gravity is beautiful"? Is it possible for one theory of physics to be "more beautiful" than another, and, if so, how exactly?

VII. MOND vs. Dark Matter, Part I

Looking at galaxies

HB: So you've been talking about Newtonian gravity and Einsteinian gravity, but what about MOND—Modified Newtonian Dynamics—that some people might have heard about? What's that about and how does that relate to this issue of dark matter?

JK: Well, one of the reasons why it took Einstein, and physicists in general, so long to come up with general relativity is that Newtonian gravity works so well. Gravity is a very feeble force, so to really see the imprint or signature of general relativity, you have to go to situations where gravity is strong, like being near a black hole or dealing with very large cosmological distances. But for most purposes, Newtonian gravity works exquisitely well; in particular, the anomalous effect of the precession of Mercury's orbit that we talked about earlier was tiny—just 43 arc seconds per century—which is why it was so hard to detect in the first place.

Now, MOND is a rather radical idea that says that Newtonian gravity should be modified, not in extreme regimes where gravity is strong, but rather in regimes where gravity is rather weak, in fact, so weak that we only probe it at large distances from the galaxy—not in the solar system environment.

It's a radical idea, I have to say, but it's a proposal that basically says that, when you get down to very low accelerations—those comparable to the Hubble expansion rate today—that occur at large distances in the galaxy, at some point Newtonian physics must be revised.

From a theoretical point of view it's a bit funny, because normally you'd expect new physics or new modifications to occur when things get strong—when you go to short distances or high accelerations. That's what happens in the context of general relativity. For MOND, however, the idea is that it's a long-distance, low-acceleration phenomenon.

It's an empirical rule, but it works amazingly well to explain the properties of galaxies. According to MOND, then, you would say, *"There's no dark matter in the galaxy, but instead there's this new law of gravity and it works beautifully at explaining what we observe about the rotation rates of galaxies."*

HB: Which was one of the major telltale signs that led to the whole idea of dark matter to begin with.

JK: Absolutely. That's one of the strongest pieces of evidence that we have. We look at how, effectively test particles—hydrogen gas—rotates around the luminous matter—the stars—and we find that, very far away from the stars, where normally you'd expect the velocity to drop as you move further away if Newtonian gravity would be correct, instead, you tend to get a velocity that's independent of distance, which is a very funny phenomenon.

Now, if that's all that was going on, it would be pretty easy to cook up some theory that would do that, but there's something else.

Galaxies are really messy objects: you form stars, they emit energy and so forth. It's a very messy process to form an actual galaxy. In the dark matter paradigm, this luminous matter is surrounded by some dark matter halo, which explains why you get this larger velocity. So at those distances you're completely dominated by the dark matter and the luminous mass represents only a very tiny fraction.

But now we turn to considering this empirical fact, which is subtle and very curious, which tells you that the luminosity of the stars in a galaxy is related tightly to the asymptotic velocity far away from the luminous matter, which has to do with the amount of dark matter. The correlation is that the luminosity goes as velocity to the fourth power.

That is a well-known, empirical law, but it's a very funny thing when you think about it. Of course you would expect that the more massive the galaxy is, the more luminous it is, and therefore, the more dark matter there should be. But why to this particular fourth power? Well, that is simply an empirical fact, so we can't argue against it.

Now, it could be that, when you do a careful enough simulation with baryons and other messy physics and dark matter, this comes out as an emergent, empirical law.

Or it could be, as is the case with MOND, that it's, in fact, a product of the theory: the theory actually *predicts* that this should be the case, which is quite interesting.

Now I'm not, myself, a direct proponent of MOND. I haven't spent years working on it or anything like that. But everybody who works with fitting rotation curves of galaxies to the data agrees that this correlation between luminosity and velocity is an amazing fit: it's one of the most successful, empirical laws in astronomy.

So, I would say that the theory of dark matter as far as we presently understand it—dark matter particles—has a long way to go before it can explain those aspects of galaxies.

Now there's *other* evidence for dark matter, which convinces me that it cannot just be MOND by itself. Once you move to the scale of clusters of galaxies, or an even larger cosmological scale, and *there*, I think, the evidence for dark matter is pretty strong—at least in the way we understand it.

HB: Let me just back up for a second.

I appreciate that you're not a die-hard MONDian, but let's just suppose for a moment that you were and you were to say to me, "*I have this equation that tells us that we have to modify our understanding of Newtonian gravity at these particular large distance scales and low accelerations.*"

I would then say to you, *"Well, that's very interesting, but where does this theory come from, exactly?"* Because it's our general understanding, in physics, that our final equations need to fit into some overall framework, some coherent picture in terms of fundamental principles that fit in with the other stuff we know.

So I would probably say something like, *"Well, your explanation seems somewhat arbitrary to me. It fits the rotation curve data of galaxies, that's very nice. It predicts this luminosity and velocity relationship, that's nicer still—and possibly quite suggestive of something deeper. But where does it **come from**? What **principles** is it based upon?"*

JK: That's exactly the point. I think, at the moment, that's what's missing. For example, how did Einstein come up with general relativity? He first developed his famous equivalence principle between inertial mass and gravitational mass. That was the guiding principle that led to general relativity—and that, together with special relativity, gives you general relativity.

We don't have that for MOND. As you said, it's kind of a cooky formula that works, but there are various hints and proposals out there. They're not well-formulated yet, but there are further avenues to explore.

For example, that law, when you're in the deeply MONDian regime, has some symmetry properties to it—it's invariant under scale transformations—and so there may be some hints of something deeper going on.

I have some wild ideas about that, too, about why this would come about.

But in short, yes, I completely agree. It looks, at the moment, just like a fitting formula—incredible predictive, but a fitting formula nonetheless.

The path that a lot of people have taken is to take that fitting formula and try to fit it in with general relativity: come up with some modification of general relativity out of which this would come out, analogous to a Newtonian limit.

However, that, to me, feels like you're trying to put a square into a round hole, it doesn't feel like it fits very nicely. There has to be a deeper principle, something that will tell us that this is really the answer. That's definitely what's missing.

Questions for Discussion:

1. Why do you think it's so atypical for theories to be modified in regimes where the relevant forces are quite weak, rather than very strong?

2. How would you describe Einstein's equivalence principle between inertial and gravitational mass?

*3. **Must** a theory of physics arise from some deeper principles? Is that, in fact, strictly necessary, or simply desirable for us based upon what we're used to?*

VIII. MOND vs. Dark Matter, Part II

Dark matter roars back

HB: Let's return to where the dark matter regime is more successful, in terms of this larger scale structure that you were just alluding to.

JK: So, where dark matter is most successful is on the larger scales. From an outsider's point of view, this is perhaps more subtle and harder to appreciate. When it comes to looking at galaxies, it's almost visceral—you can see the effects directly.

Cosmologically, on the other hand, it's more indirect, but I would say it's the firmest evidence for the existence of dark matter.

The idea is the following: we know, from the cosmic microwave background that the early universe was highly homogeneous: the density was almost the same everywhere in the universe—like, say, the surface of the ocean.

But it was not perfectly homogeneous: there were some regions that were slightly denser and others that were less dense—like ripples on the ocean. And what happened in time is that the regions that were slightly denser collapsed under the influence of gravity. Eventually, they grew so dense they led to the formation of the first stars and galaxies, and that's how everything came about.

So in principle, if you take those initial conditions—the amount of matter you have to start with, the size of the primordial inhomogeneities and the amount of time that's elapsed, you can "run the clock forwards" according to our present theoretical understanding and get to where we are today.

And it turns out that when you do that, if there's no dark matter, no galaxies form: there just hasn't been enough time for things to collapse. So, that's one piece of evidence. You need dark matter because, if you have dark matter, the gravitational wells are deeper and this process will be sped up and you'll form galaxies by today's time.

In addition, more indirectly from the microwave background patterns, there's further evidence for the existence of dark matter.

I think a theory like MOND would be hard-pressed to explain those features. That's my personal opinion. I think it works really well on a galactic scale, but because evidence for dark matter comes from these rather different places, it's very hard to imagine that a single, all-encompassing theory could explain all of these different things without dark matter. It's possible; but in my opinion it's unlikely.

HB: My understanding is that, if I'm a MONDian, or at least an old-fashioned sort of MONDian, I would say, "*No, no, there's no such thing as dark matter; this is an unnecessary hypothesis and you just have to believe in my equation and everything works out perfectly in terms of these rotation curves of galaxies and so forth.*"

But from what you were just saying—that you actually need dark matter to be able to form stars and galaxies on the right timescales—is it now the case that some people of a MONDian persuasion would now say, "*Well, you actually need both. Maybe there is this dark matter somewhere—maybe it doesn't exist at the levels some people think it does on a galactic scale, but it does exist on some other scale or some other level*"?

Because, if you don't say that—which admittedly seems to go against the basic origin and motivation of the MOND theory to begin with—then how do you address this issue that you were just discussing, which is that you need dark matter to be able to directly demonstrate how things are working on those time frames.

JK: That's right. Well, I can't speak for all the MONDians in the world, but it's a good question.

Well, what is their attitude? First of all, to me, it's sort of an extreme point of view, to say that there's absolutely no dark matter, but I think that's pretty much the view that's being espoused by that community. They need some form of dark matter, so to speak, in which they say that neutrinos are actually more massive than we think they are—which is an interesting proposal in and of itself.

So, it's not like they don't need any dark matter, but neutrinos are still part of the standard model, so it's not really "dark matter" per se...

HB: But there have been all these tests on neutrino masses.

JK: That's right. So it's barely within known, experimental bounds. But at any rate, the question is, how do they explain what I just said? Right now, I don't think there's a theory that can predict it in detail because, to explain cosmology, you need a relativistic theory, and there have been some proposals of an extension of general relativity that does that, but it has not been successful at the level that I just said.

I think it's unlikely, but they would probably just say, "*Maybe it will work out.*" There is still the possibility of a relativistic theory that would explain this phenomenon.

HB: In other words, something like, "*If we keep working on our theory, we'll eventually get a fully relativistic form of MOND which will also be able to explain this.*"

JK: That's right, yes—that's their view, I think.

Questions for Discussion:

1. What does Justin mean, exactly, when he speaks of "a fully relativistic form of MOND"? Why do you think it's so important for physicists to have such a theory?

2. Might there be additional assumptions inherent in our models that we use to "run the clock forwards" from the early universe to the present day?

IX. Why Not Both?

The liquid helium analogy

JK: My attitude is that I actually think both points of view are rather extreme, if you think about it.

Some believe only in dark matter particles, convinced that general relativity, as we know it, works perfectly fine. The other attitude, which is equally extreme in my view, says that there is no dark matter whatsoever and it's all modified gravity. But why not both?

Of course, one could cite Ockham's Razor and all those things, but really: *why not both?* There are systems, condensed matter systems that can serve as an example, here. We experience different phases of water that can coexist—ice, steam, normal water—it's all the same, underlying physics, but their individual manifestations are rather different.

One of the things that I've been thinking about is to consider the dark matter particles and the MONDian phase just as we think of superfluidity in liquid helium. Liquid helium is a very well-tested phenomenon: you cool it down and eventually it reaches a superfluid phase, but above absolute zero, there are still two coexisting phases—the superfluid phase and the phase made of the normal fluid—and they coexist.

Maybe it's something like that: there's the same underlying physics, but there are different manifestations on different scales.

Of course, it's easy for me to say those words and quite another thing to do detailed calculations.

HB: How has the general idea that these two different approaches might represent two different manifestations of the same underlying physics been received by your colleagues?

JK: I think there's a slight shift in the community. A number of people have become more open to these sorts of ideas, and I think it's been driven by various developments.

One of them is that, while dark matter experiments are becoming increasingly sensitive, they still have not discovered dark matter particles.

One of the leading candidates for dark matter particles are the so-called weakly-interacting, massive particles—so-called WIMPs—coming from supersymmetry. But now these experiments are starting to really chop down through the parameter space that would be favoured by these supersymmetric extensions of the standard model.

So at some level, there's a bit of a concern. Then there's the LHC, the Large Hadron Collider at CERN that discovered the Higgs particle. Within the next five years, or even less, we will have reached the highest energies of the LHC, and everyone is anxious to see if we will have discovered evidence for supersymmetry and/or new particles.

If they start discovering new particles at the LHC, then I'm certain that will have direct implications for dark matter. We might not be able to say, "***This** is the dark matter particle*," but the discovery of any new particles would, I'm quite certain, be directly linked to dark matter.

On the other hand, there's the pessimistic possibility that no new particles are discovered in the LHC—there are no hints, at the moment, that there will be new physics. And suppose, too, that these dark matter experiments still don't find anything. Well, then, that gives you pause.

And I think that possibility is what's been largely responsible for this shift I referred to a moment ago. It's certainly influenced me and many of my colleagues.

Meanwhile, the other development is that numerical simulations of galaxies are getting increasingly good. We cannot, in detail, simulate the messy processes of star formations, so what you have to do is make some approximations: treat the baryons essentially as a fluid and put in some parameters to ensure that you account for the energy released by stars, and so forth.

So then the question becomes, *"How well do these simulations do at getting galaxies to be the way that we observe them?"* I'm certainly not a detailed expert on the simulations—I don't do them myself—but my understanding is that the many parameters we're dealing with now are parameterizing our ignorance about what the actually messy, baryonic physics is—you have many knobs to tweak. It's not clear whether or not it's working out, and I think the jury is still out on that.

Those are the two developments that I think are giving some people, at least me, a lot of pause. Meanwhile, the evidence for dark matter from the other side—from the larger scale—is increasingly good.

So, from my point of view, I think this hybrid viewpoint is increasingly worth pursuing: it's guided by the data, in some sense. As you said, it's still in need of a good, compelling, theoretical explanation, but it's certainly worth putting time and effort into.

Questions for Discussion:

1. Why do you think Justin mentions "Ockham's Razor"? What is that and how has it been used in the development of theories of physics?

2. What do you think Justin means, exactly, when he talks about "parameterizing our ignorance"? To what extent is this a standard approach to a deeper level of scientific understanding?

3. What role do analogies, such as Justin's invocation of liquid helium, play in the development of our understanding of theories of nature? To what extent can we objectively distinguish between good, or potentially relevant, analogies and poor ones?

X. Dark Energy

Vacuum energy and the cosmological constant problem

HB: Let's tack to dark energy now. Once again, there's the obvious question to ask. You cosmologists now tell us that there's this mysterious force out there which is causing the universe to accelerate in its expansion, but how certain are we of this? How do we know that this is actually going on?

JK: Again, it's rather indirect in some sense through gravity. The first thing to say is that, from an outsider's point of view, the most surprising fact about cosmology is the fact that the universe is expanding in the first place. That's the most remarkable thing: gravity is attractive, so why is the universe expanding? The most naive thing would be to expect that the universe would be collapsing under its own gravity. Of course, the honest answer from cosmologists is that we don't know, because that traces all the way back to the Big Bang.

We don't know what happened at the Big Bang, but that's really where the expansion started out, so the key really lies in the initial conditions and it's been expanding ever since. Now what we would expect from gravity, at least, is that it would pull stuff in, the expansion should slow down.

And that's what seemed to have happened for most of the history of the universe, the expansion was slowing down. But then the surprise comes in that, about seven billion years ago, the expansion of the universe started accelerating. Ordinary matter could not do this, because ordinary matter would make gravity attractive. There's some other, funny, form of energy that is driving this acceleration to occur, and that's one way that we can tell that there has to be this form of dark energy.

The fact is that we can observe, we can measure and infer, the expansion rate of the universe; that's a pretty direct, kinematical test: you can measure the expansion of the universe as a function of time and you can see that it's been speeding up for the last seven billion years or so. So, if you believe in Einstein's theory of gravity, then there has to be some missing form of energy.

And this missing energy can't be dark matter, because dark matter would clump and make the universe slow down. So, whatever it is, it's not collapsed in galaxies, as far as we can tell, it's more or less homogeneous, and the simplest explanation would be that it's vacuum energy—the energy of the vacuum from quantum physics. That's the leading explanation, but this has problems as well.

Regardless, that's probably the simplest explanation we have right now as to what dark energy is.

HB: Let's talk a bit about those problems you just mentioned, starting with the disparity between what a particle physicist might calculate for this vacuum energy and what we actually observe.

JK: Well, yes: this is the worst failure of theoretical physics.

First off, when we say "vacuum energy", this is the energy of quantum fluctuations in the vacuum. In quantum mechanics, there's no perfect vacuum: things always pop in and out of existence in the vacuum, and we can calculate this energy of the vacuum from the particles that we know.

Every particle in the universe is described by a field—just like the electric field, for example—and we can compute what these fluctuations would contribute to this average vacuum energy. But when you do the calculations, you find that you get the wrong answer by something like 120 orders of magnitude. It's the worst failure; no theoretical physicist would like to talk about this.

HB: Well, just to highlight things, that's a *colossally* bad result: 120 orders of magnitude means a one followed by a 120 zeros. That's not just wrong, that's remarkably, outstandingly wrong.

JK: That's right—it's a big problem. Now, what does it mean?

One thing to say, of course, is that we cannot calculate the precise answer of the vacuum energy, because we don't know all the particles that are there or all the contributions to this vacuum energy, but we can calculate the contributions of those that we do know, and those are the ones that give you the wrong answer.

There are two attitudes really. One is that we get the wrong answer, but, in fact, there is another big contribution, which we can't account for, which cancels this at an amazing precision of one hundred and twenty decimal places. That is essentially the conventional answer right now—that this cancellation occurs.

Now that makes virtually everybody very uncomfortable, because we would like our theories to be somehow natural and generic; you shouldn't have to tweak things to get the universe to work out.

The attitude in the field right now—and I'm talking about this conventional view—is that this cancellation may not happen everywhere, that we happen to live in a special place where this cancellation occurs between these different components.

If the cancellation didn't occur, the universe would be expanding or collapsing so fast that we wouldn't exist. That's an argument that's been proposed.

I don't particularly find that argument very compelling. I think in the history of physics, whenever we have these kinds of issues where our calculations tell us that something is wrong, it's usually a sign that there is new physics going on that we're about to discover that will explain, naturally, what's going on.

So, that's the more hopeful attitude: that there has to be something beyond what we currently understand. It may still be vacuum energy the way I just described it, but there's something that makes it appear weak, and for that there has to be new physics.

HB: Well, these are very complicated issues and they've been around for quite some time now, and I don't want to trivialize them or say that people who believe one thing are naive or silly or anything, because I certainly don't believe that.

But still: if I'm some guy on the street, and you come to me and you say, "*Look, I'm a scientist and I have the answers to how things work in this particular regime,*"—let's not even talk about cosmology, let's talk about something else, say, how my coffee cup slides across the table.

And I say "*Okay, great, I'd love to hear your theory.*" And then you explain, "*Well, according to my theory, your coffee cup is going to go upside down and it's going to go at the speed of light between here and there; it's going to turn into a frog over there, and then magically reappear where you placed it, just as if you had been pushing it along at this particular rate.*"

JK: That sounds a lot like the talks I hear at conferences these days.

HB: Right. And, I would say to you, "*That's a very interesting theory, let's test it,*" and as I'm looking at the coffee cup, I would think to myself, "*Well, hang on, it doesn't do any of the things that you just proposed it does.*" To which you would reply, "*Ah, yes, well, I have an answer for that, you see, because what's happening is that there's another effect that I can't quite predict that cancels it doing all of those things; and so what we see, when we take everything into consideration, is what actually seems to be happening here in front of us.*"

So I would naturally say, "*Well, Justin, I'm pretty suspicious of your theory,*" because it seems to me that you could say just about **anything** to account for the different measurement I am experiencing from what you had said was supposed to happen.

Again—with respect, because I well appreciate that these are very complicated issues and I'm trivializing them to make a point—but I just think that when you have a prediction that jars with the experiment to the tune of a 120 orders of magnitude, to be able to say that there's something else that happens to just come along and cancel that—well, you seem to really be reaching for something.

JK: Yes. I think that kind of scepticism is really healthy because, in any scientific endeavour, we get used to certain assumptions, certain approaches, and we really have to step back and realize just how crazy this all is and truthfully ask ourselves, "*Are we missing an opportunity by sweeping under the rug these kinds of real issues?*"

It's one of those things where it's lacking a better explanation, and we have to be honest about it. As a theorist, I can be honest that there's nothing, at the moment, that can compete with this outrageous explanation. There are some ideas, but nothing at any significant level of mathematical detail.

As you know, we don't just ponder crazy ideas like this: we have mathematical rules to write them down, we have a certain set of guidelines that we believe in. So, you can't just do whatever you want. And when I go work on my theories this afternoon, I quickly find what I can do is very limited; and that's why people haven't found a competing alternative.

There are certainly ideas out there and many people are working on this, myself included.

So here's one idea, just for example. When we say "the cosmological constant," or the "vacuum energy," this form of energy is the most homogeneous form of energy that we've *ever* probed: it's homogeneous both in space and in time—it doesn't change in time.

So the one thing you have to worry about is that it means that you're probing, through gravity, the largest possible scales in space and in time, because it's perfectly homogeneous. But do we really know that gravity works the way we think it does on those huge scales?

Maybe, in fact, the following could be true: that the vacuum energy is extremely large—this 120 orders of magnitude 'wrong answer 'is the correct answer—but the fact of how it gravitates is different than we assumed.

Maybe it actually gravitates 120 orders of magnitude more weakly than you and I gravitate. We've tested gravity on small scales, we know how we gravitate, we know all that but, when you get to these extremely large distances, do we *really* know how vacuum energy gravitates?

That's one idea that I've explored, and one for which I can actually do some mathematics to back it up. Again, in terms of aesthetics and theoretical control, it's under the same level of your "frog to cup" theory, but it's at least a little more precise, it's not just words.

Questions for Discussion:

1. Are theories in cosmology falsifiable in the way that theories in other branches of physics might be? How might this distinction further complicate the "cosmological constant problem"?

2. To what extent is Justin's speculation on different types of "gravitational regimes" for the cosmological constant problem structurally similar to the views espoused by those who subscribe to MOND?

XI. Personal vs. Professional

The scientist as childlike iconoclast

HB: We've just talked about two major conundrums, two major unsolved difficulties—dark matter and dark energy. Might it be the case that these things are actually related at some level?

JK: Absolutely. That would be the most compelling answer: that somehow these vastly different phenomena somehow have a common origin, but that's not how most cosmologists think about it.

There's a funny discrepancy between what people talk about at the conferences and the formal talks. We give talks that are rather conservative, because that's how we're trained. That's a good trait to have, actually, to be conservative: we're not completely wild.

But at the end of conferences, sometimes somebody will want to have some fun and say, "*Let's have a survey, let's write down some options for what we think dark energy is*".

And it's always startling that the way people vote on these surveys is far from conservative: they vote for the most radical option. Now, why is this? Well, maybe it's because we are hopeful: we don't want to put our names on a paper that's crazy, but deep in our heart, we hope that it's a radical answer so we have more stuff to do for the next 20 years. Or it could be a real belief: we don't have an idea for how to actually explain it and write papers about it, but we actually believe that it's true.

It depends how we phrase the question, but if you ask, *Would most cosmologists think that linking these things is a radical idea?*

I think so, but it's difficult to tell. But for me, that's my hope: that these phenomena **are** connected.

In fact, it would be great if *all* of these phenomena - this MONDian modified gravity on galactic scales, the dark matter and the dark energy are somehow—again, this is the dream—different phases of the same, underlying stuff that we have to understand.

HB: Depending on how one interprets it, one could make an argument that there is evidence for this type of thinking working very successfully in physics over the years and even centuries— seemingly disconnected phenomena, disconnected approaches that were later seen to be manifestations of the same thing— this has happened over and over again. To some extent, people have accused physicists of taking that approach to the exclusion of all other approaches, insisting that everything has to be manifestations of the same thing. But I certainly think there is a historical argument that this is something worth approaching or worth thinking about at some level.

Listening to you talk about your anecdotal experience of the jarring distinction between what researchers think in terms of their own personal opinions manifested in an informal vote at the end of conferences as opposed to their papers, it's almost as if, for a moment, these researchers are little kids again.

They're thinking about the same sort of thing that got them into science and they're letting their imaginations run wild, but they realize, of course, that they can't do that when they're being professional scientists and when they're worried about going through tenure and promotions and all the rest of that.

But at some level you're hitting the core aspect of why they went into science to begin with—it's like a window on their own past.

JK: That's right. At some level, I had that experience with the gravitational wave "discovery", which turned out to be a non-discovery.

It was interesting, because this claimed discovery of primordial gravitational waves appeared to rule out these ideas of the Ekpyrotic Universe I was describing earlier—the theory that me and my colleagues had worked on—and confirmed the consensus theory, which is inflation.

At some level I should have felt disappointed, but at the same time, I was so excited for the field that this was happening, that we had this new piece of evidence about the very early universe, that it trumped my own personal interest. As it happens, it turned out that this wasn't a discovery after all, but it was an interesting personal experience to go through that.

I think that most people in the field feel that way. Ultimately, as you said, we want the greater good of the field: we want to be excited about things, we want nature to surprise us. I think that's an intrinsic feeling for any scientist.

Questions for Discussion:

1. How do you think most physicists would react to the prospect of a "Theory of Everything" being discovered?

2. Do you think that we are doing a sufficiently good job of encouraging scientists to "think outside the box" and develop theories that are well outside the mainstream? How, concretely, might that be more encouraged, while still retaining the integrity of the enterprise?

XII. Revolutionary Rumblings

Cosmology's Golden Age

HB: I'm going to get you to lay it on the line for me now, Justin. You've given some hints as to what *might* happen in the future, but I want you to tell me what you *think* we're actually going to discover in the next 10 or 20 years.

JK: OK. I won't hedge my bets. I'll tell you what I think are the two scenarios, and then I'll tell you which one I prefer.

The first route is the one that we've been following: that dark energy is vacuum energy, that dark matter results from these weakly interacting particles—whether we discover them directly or not isn't essential, that's what the data will actually tell us. That's route number one.

Route number two is a radical one: a revolution in our understanding of gravity and in our understanding of what this dark stuff is. Why do I only see these two options? Well, I don't think it's going to be a tweak over what we understand. It would be deeply unsatisfactory just to tweak.

HB: Also, people have been trying to tweak things for so long that it considerably increases one's scepticism that it's tweakable at all.

JK: That's right, "*Why this tweak and not some other tweak?*" Yes.

So it's more than a hope, actually. My *belief*, I think, is that it will be route two. I think that the universe can still surprise us. In fact, this is not just a statement about 20 years down the road, I think that this is a critical juncture, right now, this particular period.

I believe that we're at a juncture where these two roads will diverge and it will come from increasingly precise data, increasingly precise simulations. It might come from the LHC, it might come from something else, but these dark matter experiments, at some point, reach a limitation, and I think we will know for sure, in the next few years, which route we're going towards; and I think we're moving in that direction.

HB: Very good. And given that that's your gut feeling, but also understanding the field and recognizing the spectrum of possibilities, what advice would you give young people who are contemplating entering this field? Is there anything specific that you'd recommend?

JK: I think they should definitely jump in. I think it's the best time ever to jump into this field, precisely because we're about to enter this revolution. I think we're about to experience the same sort of revolution that happened at the beginning of the 20th century with the invention of relativity and quantum mechanics.

So, if you're a young person, I think it's the perfect time to jump into this field and contribute something original. The problem with someone like myself who's been in the field for a long time is that we get blasé. We think, *"Oh no, this idea will go away"* and that sort of thing, but young people don't have that, and I think that's really helpful and refreshing.

So, I would say, *"Don't be afraid to come in, propose new ideas, think outside the box and think about alternative systems that we've thought about in history that connect these different phenomena."* We need fresh blood.

HB: Sounds great. I really had a great time talking to you, Justin. Anything else? Did we miss anything, you think?

JK: No. That was great, thanks.

Questions for Discussion:

1. Would you consider studying cosmology professionally if you could go back in time and become a student again?

2. Are we doing a good enough job communicating the outstanding mysteries of nature to high school students?

3. Should more public money be spent on subjects like cosmology even if such research doesn't have any strong likelihood of yielding practical applications?

4. Is it possible that the universe is simply too complicated for us to understand?

Continuing the Conversation

Additional Ideas Roadshow conversations not offered in this collection that the reader might enjoy include *The Power of Principles: Physics Revealed* with Institute for Advanced Study particle physicist **Nima Arkani-Hamed**, *Harnessing the Sun* with Imperial College solar cell physicist **Jenny Nelson** and *Science and Pseudoscience* with Princeton University historian of science **Michael Gordin**.

Astrophysical Wonders

A conversation with Scott Tremaine

Introduction

On Butterflies and Fish

In many ways, the march of scientific progress can be viewed as an unrelenting peeling away of our innate sense of uniqueness. In the last few hundred years, we've gone from regarding ourselves as a divinely-favoured species placed squarely at the centre of everything, to a randomly evolved concatenation of DNA on an unremarkable planet orbiting an average type of star in a not terribly noteworthy position of a fairly run-of-the-mill type of spiral galaxy in a gargantuan, and thoroughly indifferent, universe.

Such a precipitous descent from the privileged to the mundane is humbling, to say the least, but in what might be somewhat sceptically regarded as a desperate attempt to diligently reassert a sense of grandiosity to the situation, many have now elevated our reluctantly-concluded averageness to nothing less than a cosmological statute: the Copernican principle—sometimes called the mediocrity principle—inverts the argument to claim that everything about our situation is overwhelmingly, necessarily, common.

Just as grabbing someone off the street at random will result, in nine out of ten cases, with coming face to face with a right-handed person, one should naturally expect that our planet, star, solar system, galaxy and particular region of the universe—taken, as it were, at random out of all the possible places we might have found ourselves—should instead be, well, pretty much the same as just about everywhere else.

But as fond as we are of postulating principles, modern science relies even more strongly on measurement. The philosophical pendulum might have swung from one end to the other, but now that we have the tools to carefully examine plentiful numbers of planets and planetary systems outside of our solar system, what does the experimental evidence actually tell us?

Scott Tremaine, Richard Black Professor of Astrophysics at the Institute for Advanced Study in Princeton, New Jersey, and an internationally renowned expert in both galactic-scale and planetary-scale astronomy, was an ideal person to ask. So, does the mediocrity principle hold sway, or does our solar system veer surprisingly towards the left-handed?

> *"Over the last two decades, we've discovered hundreds, probably thousands, of planetary systems orbiting other stars. We now know, then, that planets are common. This wasn't at all obvious. In fact, in the late 19th century and the first half of the 20th century, the standard model for planet formation involved a very rare, unusual event: the close passage of two stars, which was likely to have only happened a handful of times in the galaxy. So, in that model, the solar system was this extraordinarily unique and unusual configuration that you practically never saw.*
>
> *"That model had already been superseded in the 1960s, but we now know for certain that it's wrong, because if you look at a typical star, even with our limited abilities to detect planetary systems around other stars, probably a significant fraction—10%, 20%, 30%, depending on how you define it—have planets around them."*

Well, that seems to settle it. After all, if huge numbers of stars come with planets orbiting them, then that seems to give us a definite push towards garden-variety status.

But then, there's this:

> *"Most systems we've seen don't look like the solar system. Many of them have giant planets that are much closer to the host star*

compared to our own giant planets. In fact, many of them have
planets that are much closer to the host star even than Mercury,
the innermost planet in our own system.

"We don't have a good theory for how those formed, and we
don't have a good theory for why they're different from our own
planetary system. One of the reasons for the difference, of course,
is that it would be quite hard to detect our own planetary system.

"If we were sitting on a planet orbiting a star a few light-years
away and were conducting the same planetary surveys with our
current technology, we would just barely be able to see Jupiter.
And we would be a little puzzled, because Jupiter is in a much
more circular orbit than most of the analogs to Jupiter around
other stars.

"So, astrophysicists sitting on this hypothetical planet ten light-
*years away would probably have said, '**That system looks a little***
unusual because the planet that we see is a lot more circular
***than most of the planets that we see.**' "*

Score one for uniqueness, then. Perhaps we're not so common,
after all?

The short answer appears to be that it's simply too early to tell.
Given the fact that modern approaches to exoplanet detection are
only a few decades old, it could well be that the statistics will level
out as time, in tandem with detection technology, evolves. But then
again they might not. Deciding between the two possibilities is
simply an empirical question.

For Scott, however, there is a vital lesson to be learned here about
transcending personal biases that extends far beyond the statistics
of one particular solar system, however near and dear it may be to
our hearts.

The goal of any scientific enterprise is to understand phenomena
in the most general possible terms in accordance with the laws of

nature. But in our quest to generalize, we often unthinkingly limit ourselves to our own particular experiences.

"The analogy I sometimes use is: suppose Darwin were developing the theory of evolution and all he'd ever seen was a butterfly. He might get the general idea of the theory of evolution right, but he probably would have put some things in it to explain why evolution could only produce things with thin wings of a certain size that fluttered around from plant to plant. We now know that, in addition to butterflies, there are lions and tigers and fish and birds; and that doesn't mean that the basic idea of evolution would be different, but you have a much better idea—much better feedback—on how it must actually have worked, from the fact that you see this tremendous variety of systems."

But, what if you don't have the good fortune to be presented with such variety in the first place? What if we're stuck in the dark, like Darwin and just his butterfly? What do we do then? That, says Scott, is the core intellectual challenge of astrophysics, what keeps him coming in to work each day with a smile on his face.

"This is an example of one of the things I enjoy about astrophysics: its detective nature—having to figure out to what extent you've been assuming that everything is the same as the things we've already seen. You have to have the imagination to ask if there are other things that are allowed by the laws of physics that we haven't detected which might be quite different; and, if so, should we have seen them already, and are there techniques for detecting them?"

You also have to learn from past mistakes and lost opportunities. As it happens, the burgeoning field of exoplanet astronomy has some lessons to teach us there, too.

*"The irony, I think, is not so much that the technology is now good enough to detect exoplanets, it's that these planets probably **could have been detected** much earlier, but the planets that you can see by these techniques are those that are quite close to the host star;*

and all the people who were looking for extra-solar planets in the 60s, 70s and 80s, when the technology was already good enough to detect them, thought that we should be looking for systems like our own solar system, in which case neither of these two methods would have worked.

"Because we'd only seen one planetary system, then, the community not unusually made the assumption that all the other ones should look the same. As a result, nobody really took seriously the idea that you could use these techniques to deduce planets.

"Again, it comes down to the analogy of Darwin with the butterfly—no matter how well you understand evolution, if all you've seen is a butterfly, you're not going to predict the existence of fish."

We are all, to some extent, naturally limited by our pasts, unable to leverage first-hand encounters with the entire variety of potential systems. But the successful scientist can overcome the limitations of her past through a conscious act of creativity: harnessing her imagination to envision what might lie beyond her experiences, thoughtfully designing both theories and experiments that can capture the possible.

Such critical thinking skills may or may not be unique in the universe. But if we don't exploit ours to the fullest, we'll surely never find out.

The Conversation

I. Personal Reflections

Astrophysical origins and research-administration balance

HB: I'd like to go start at the beginning of your own personal scientific history. Was astronomy a passion of yours from a very early age?

ST: No. In fact, at a young age, I was pretty convinced that astronomy wasn't very interesting. I first got interested in physics largely because I realized, when I was in high school, that for most of the other courses, I had to memorize a whole bunch of facts, which I was terrible at—whereas in physics, once you understood the basic concepts, you really didn't have to memorize anything. So, I got interested in physics because it was easier and less boring than, say, trying to memorize the capitals of all the Soviet Republics.

HB: Okay, I'd like to push that a little further but first let me explore your lack of interest in astronomy per se. Why was that something particularly *uninteresting* to you?

ST: Well, I think astronomy, as typically presented in popular books, is a set of facts, similar to stamp collecting. You're presented with constellations, or people tell you that stars have a particular form, but there's no logical argument to it that you can follow. It was only when I started learning about astronomy in the context of physics that I realized that a better way to think of it is that astronomy is taking the laws of physics that you learn in the laboratory and applying them to what's going on in the heavens.

In fact, the idea that you can do that is actually pretty recent: Aristotle thought that the material in the heavens was of a completely different nature to the material on earth. The notion that you could measure something in the laboratory and then apply it to stars or the sun or galaxies is something relatively recent—from the last couple of centuries.

HB: Right. Getting back to your story: so you moved into physics initially out of a sense of indolence, shall we say—to avoid all of this other mindless memorization you were faced with—but were you aware, at the high school level, of this sense of overarching principles? Clearly you understood the obvious practical implications—because you didn't have to memorize a bunch of stuff, you could derive what was needed in accordance with the underlying principles—but was there an aesthetic appeal to you, at the time, this way of looking at the world from a few, essential principles? Did that start resonating with you, even back then, in an intellectual way?

ST: Yes, I think that first started to be a point of view that I understood in high school—and, of course, much more strongly at university.

HB: Did you have a particularly influential high school teacher?

ST: I've had a number of influential teachers. My high school teacher was influential because he let me fool around in the equipment room after school. I'm sure that's something you wouldn't be allowed to do today, but I had a great time hooking up lights and Van der Graaff generators and that sort of thing. I think the freedom to play around with equipment was a real asset.

HB: Then you went off to university, and presumably your interest deepened. How did the astrophysics component arise?

ST: Well, I went to McMaster University, which at the time, didn't have any research focus on astrophysics, but one of the professors taught a second-year course on astrophysics and I began to realize that all of these techniques that I had learned in other contexts— in the context of laboratory experiments—could actually be applied to understanding all these facts that I'd learned as a kid about astronomy, which made them much more interesting.

The other appeal of astrophysics is that, in a certain sense, it's more like a detective story than other branches of physics, because in other branches of physics if you have something you don't understand, you try to design an experiment that's going to allow you to understand it.

In astrophysics, however, you often *can't* design any experiments. You have a much more incomplete set of clues; and, like Sherlock Holmes, you're trying to deduce what must have happened from partial evidence. Also, like in Sherlock Holmes, the game of knowing *when* you've got the answer—because one answer is so much more compelling or simple or beautiful than all the other possibilities—is often more sophisticated in astrophysics than in other branches of physics.

HB: So, there was a real, intellectual resonance with the process of discovery.

ST: Well, the process of discovery is quite different from other branches of physics. In some ways, it's better because, as I say, it's more of an intellectual puzzle because the information is so incomplete, but in some ways it's worse, because if you don't know the answer, you may not find out what it is for another 10 or 20 years—or maybe even never—as the quality of the observations improves.

HB: I'm going to make a small diversion now to talk about something I don't usually discuss, perhaps prompted by our past shared experiences. I'll get back to all the astrophysics stuff shortly, but I'd like to now explore your involvement in the administrative side of the scientific and academic world, which has long struck me as quite singular.

You're one of the very few people I know who has been a very successful, practicing scientist while also seeming to embrace the opportunity to do considerable administrative work—you built an institute for theoretical astrophysics in Toronto, you were chair of the department at Princeton, you've sat on many advisory boards—including chairing one I cajoled you into serving on when our institute was at a formative stage of development—you seem to have this boundless capacity to be able to participate in substantial, impactful administrative work at the same time as doing frontline scientific research, which has long puzzled me. It's almost as if you enjoy it. Is that true?

ST: Well, it is certainly true that most physicists wouldn't like to admit that they enjoy administration. I think that many people find that they can do administration for a limited period of time, but if you take on a heavy administrative job—like being a dean or a university president or directing a large institute or even a large collaboration—you can do that for a few years, but if you do it for more than 5 or 10 years, you find that it's very hard to get back to doing science effectively.

I think one of the reasons that people are nervous about this is that they're very much aware that they've seen many of their colleagues who take on administrative roles feel remorseful because they can't continue to do research and teaching.

My own view is that the selection process for research is almost entirely based on solely one's research ability; and one of the difficulties that universities have is that they're trying to pick their administrators from a set of people who weren't selected for any administrative abilities. So, it's just good fortune, really, if you happen to find somebody who's good at both.

HB: In fact, I think it's even worse than that. I would argue that being proficient at research, at least statistically, precludes one's ability to take a larger view and thus be a good general administrator.

ST: Well, you might say that; but I couldn't possibly comment. I think, though, that I've been fortunate in the administrative roles that I've had. There are some administrative roles that have a lot of responsibility and no power, or a lot of responsibility and no budget; and I was fortunate, both in the role I had as a director of an institute (the Canadian Institute for Theoretical Astrophysics) and as a department chair (at Princeton University), that those were two jobs where you had a tremendous amount of latitude, control and a large discretionary budget.

That makes things a lot easier, and it makes them a lot more fun. It's also true that, in both cases, there was a strong belief from the people who were running the place that if you let your research go, then you weren't doing your job, properly—that is, a big part of your job was to continue to do research.

In many academic-administrative jobs, either the load that you have to carry is too large to allow you to do that, or the administration just wants to see that the forms are filled out and doesn't much care about anything else. A significant concern is that the trend in most universities has been to leave the department chairs with less and less power over the years, which means that the job of being a department chair has become less one of studying the intellectual directions for the department and more one of being a minor bureaucrat.

HB: Just operationally—to close off my diversion—when you were in these major, administrative roles, did you spend a day doing science and a day doing administration? Did you cut it down to four hours a day here and three hours a day there? How did you manage your time, and did you find that your ability to manage it changed over time?

ST: I know people who do split things up that way: there are even people who have two offices and go to one to do their administrative work and the other to do their research work. I never found it was very practical to do that because, if you've got somebody you have to meet with and they can only meet at 3 in the afternoon, you can't really tell them, "*Well, I only do administration in the morning.*" I also found that having commitments to do research—having graduate students, going to seminars and so forth—was important to force me to stop doing the administrative work. Sometimes the best is the enemy of the good, as they say—it was better just to get something out that was adequate and then go on and do something else.

Questions for Discussion:

1. Do you agree with Scott that astronomy, as typically presented in popular books, is a set of facts, similar to stamp collecting? How might this be improved upon? To what extent are other scientific activities also portrayed this way?

*2. To what extent are today's students actively discouraged from engaging in hands-on laboratory work and what can be done to correct that? Different aspects of this issue are also discussed in two other Ideas Roadshow conversations: Chapter 10 of **Indiana Steinhardt and the Quest for Quasicrystals** with Princeton University physicist Paul Steinhardt and Chapter 5 of **Ocean Enlightenment** with marine biologist and environmentalist Edie Widder.*

3. What do you think Howard is implying, exactly, by his comment, "would argue that being proficient at research, at least statistically, precludes one's ability to take a larger view and thus be a good general administrator"?

II. Exoplanetary Insights

Looking beyond to assess our uniqueness

HB: Moving back into astrophysics, now: you've done a great deal of work at both planetary scales and galactic scales and I'd like to talk about both of those, as well as the cosmological implications of the latter.

It seems to me that it's somewhat unique to be so involved in both of these areas simultaneously. Am I completely wrong there, or is it unusual to have a foot in both camps so strongly?

ST: It's unusual, but by no means unique. I think that one of the reasons that I've tended to do that is that a lot of my research has been focused on dynamics, basically on the gravitational n-body problem and its various manifestations: you take a large number of masses, you let them interact via gravity, and then you observe what they do.

That's a scale-invariant problem—so if you understand what's going on in a planetary system, then in principle, you can understand exactly the same processes in a galaxy or a much larger-scale system. One of the things I find interesting is trying to pursue those analogies, asking, *Here's a certain phenomenon that takes place on planetary scales—is there an analogue to that on much larger scales? If not, why not? And if so, what can you learn from one that will tell you something about the other?*

HB: Presumably there are times when you can't actually carry that over from one domain to the other, when there are some structural impediments at play.

ST: Absolutely. For example, the earth has gone around the sun five billion times, while the sun has gone around the centre of the galaxy a few hundred times, which means that the processes that you have to think about in the two different cases can be completely different, as can the level at which you can prove things.

In some respects galaxies are easier, in some respects planetary systems are easier, but it's always worth thinking about *why* they're different, given that it's the same, basic problem.

HB: Let's focus on our solar system for the next few moments. Give me a sense of what we know, writ large, about our solar system in terms of how it came to be, its present configuration and its stability.

ST: The hardest of those questions is the question of how it came to be. It's gotten a lot easier—we've learned a lot more about this question in the last couple of decades—but it used to be that this was the only planetary system that we knew about, and so all of the theoretical work was biased by the fact that you couldn't tell whether a particular property of the system was fundamental or accidental.

For example, the largest planets in the solar system—Jupiter, Saturn, Uranus and Neptune—are all outside the small planets like Earth. They're all at distances from the sun that are somewhere between 5 and 30 times as large as the distance of the earth to the sun. Is that an accident? Do giant planets like that *have* to form out at those distances? Or is it just a peculiar feature of the solar system?

Over the last two decades, we've discovered hundreds, probably thousands, of planetary systems orbiting other stars, and you can begin to look at the statistics and get a much better idea of what the generic features of planetary systems are. Then the question becomes whether you can explain those generic systems.

The analogy I sometimes use is: suppose Darwin were developing the theory of evolution and all he'd ever seen was a butterfly. He might get the general *idea* of the theory of evolution right, but he probably would have put some things in it to explain why evolution could only produce things with thin wings of a certain size that fluttered around from plant to plant. We now know that, in addition to butterflies, there are lions and tigers and fish and birds—and that doesn't mean that the basic idea of evolution would be different, but you have a much better idea, much better feedback, on how it must actually have worked from the fact that you see this tremendous variety of systems.

HB: So, what have we learned exactly, now that we have a better idea of the variety of these sorts of systems?

ST: Well, we now know that planets are common. This wasn't at all obvious. In fact, in the late 19th century and the first half of the 20th century, the standard model for planet formation involved a very rare, unusual event—the close passage of two stars, which was likely to have only happened a handful of times in the galaxy. So, in that model, our solar system was this extraordinarily unique and unusual configuration that you practically never saw.

HB: Do you think that model arose because there was some a priori motivation to establish our "uniqueness", or was that just an objective, best guess at what happens?

ST: I don't know. It is certainly true that there have been a number of instances where otherwise very respectable physicists have been influenced in their views of the solar system by religious beliefs, but I don't know if that was the case in this particular example.

   At any rate, that model had already been superseded in the 1960s, but we now know for certain that it's wrong, because if you look at a typical star, even with our limited abilities to detect planetary systems around other stars, probably a significant fraction—10%, 20%, 30%, depending on how you define it—have planets around them.

So we now know that planet formation, however it works, is an extremely common process. We also know that most systems we've seen *don't* look like the solar system. Many of them have giant planets that are much closer to the host star compared to our own giant planets. In fact, many of them have planets that are much closer to the host star even than Mercury, the innermost planet in our own system.

What we don't know, first of all, is *how* those formed. We don't have a good theory for how they formed, and we don't have a good theory for why they're different from our own planetary system. One of the reasons for the difference, of course, is that it would be quite hard to detect our own planetary system.

If we were sitting on a planet orbiting a star a few light-years away and were conducting the same planetary surveys with our current technology, we would just barely be able to see Jupiter. And we would be a little puzzled, because there's a range of eccentricities in the extra-solar planets—deviations of the orbits from circular ones—and Jupiter is in a much more circular orbit than most of the analogues to Jupiter around other stars.

So, astrophysicists sitting on this hypothetical planet ten light years away would probably have said, "*That system looks a little unusual because the planet that we see is a lot more circular than most of the planets that we see.*"

Now, maybe that will go away as the statistics improve, maybe it's just an unusual coincidence that we shouldn't pay any attention to—we'll have to wait and see. But at the moment, most of the systems that we know of don't look anything like the solar system, and we're not quite sure if that's because of our limited observational ability or because our solar system is special in some way.

HB: Let's talk a little bit about the observational techniques and how people find these exoplanets: what, specifically, they're looking for and how do they find it.

ST: The problem is not so much that planets are faint, but that stars are bright. If we had a planet, like Jupiter, that was just floating in interstellar space and was as bright as Jupiter, we could easily detect it with modern telescopes. The problem is that the planets are very close to the host star, and the host star is much brighter. If you looked at the solar system from outside, the sun would be a billion times brighter than Jupiter and 100 billion times brighter than the earth, and they're so close that the planets are lost in the glare from the star.

An analogy that's sometimes used is to imagine that you're up in space a few thousand kilometres away looking at a lighthouse, and there's a firefly buzzing around in the lighthouse: the firefly is the planet, the lighthouse is the star, and the problem detecting the firefly is roughly equivalent to detecting the planet. What that means is that you can only find them through indirect techniques; and there have been two that have been, by far, the most productive to date.

One is that, as the planet orbits the star, the star undergoes a much smaller orbit, influenced by the planet, around the centre of mass between the planet and the star. That's a very small motion, but it means that the star moves towards you and away from you at speeds of a few meters per second, and that oscillation can be detected from the Doppler shift of the spectral lines in the star if you have a very accurate spectrograph. The technology for developing those spectrographs has improved dramatically over the last couple of decades and they're now able to detect motions of a meter per second or so. So you simply look for periodic oscillations in the spectral lines of the star, and from that you can indirectly deduce the presence of a planet.

The second is that, in some small fraction of the systems, the planetary orbit is aligned with the line of sight to the star, and that means that every time the planet goes in front of the star there is a temporary dip, typically lasting a fraction of an hour or so, which is occurring because the planet is blocking out some of the light from the star. That can sometimes be confused with other signals—fluctuations due to twinkling of stars, for example—but it's periodic (it keeps repeating), so eventually you can be confident that you can uniquely distinguish the effect.

The irony, I think, is not so much that the technology is now good enough to detect this, it's that planets probably *could have been* detected much earlier but the planets that you can see by these techniques are planets that are quite close to the host star, and all the people who were looking for extra-solar planets in the 60s, 70s and 80s, when the technology was already good enough to detect them, thought that we should be looking for systems like our own solar system, in which case neither of these two methods would have worked.

Because we'd only seen one planetary system, then, the community not unusually made the assumption that all the other ones should look the same. As a result, nobody really took seriously the idea that you could use these techniques to deduce planets. Again, it comes down to the analogy of Darwin with the butterfly—no matter how well you understand evolution, if all you've seen is a butterfly, you're not going to predict the existence of fish.

HB: Right. It's a play on this "uniqueness issue", but rather than necessarily assuming that we're the only ones, you're looking for copies of exactly what we are elsewhere.

ST: Yes. Here, again, is an example of one of the things that I enjoy about astrophysics: its detective nature—having to figure out to what extent you've been assuming that everything is the same as the things we've already seen. You have to have the imagination to ask if there are other things that are allowed by the laws of physics that we haven't detected which might be quite different; and, if so, should we have seen them already, and are there techniques for detecting them?

Questions for Discussion:

1. Do you find Scott's analogy of Darwin and the butterfly helpful? Which other areas of science do you think the same analogy might be fruitfully applied to?

2. What impact do you think religious and cultural beliefs had on the development of scientific theories of our solar system?

*3. How does the distinct nature of the process of discovery of astrophysics, as outlined by Scott in these first two chapters and symbolized by his analogy of Darwin and the butterfly, relate to the notion of Popperian falsifiability? Those particularly interested in this issue are referred to Chapter 8 of the Ideas Roadshow conversation **Science and Pseudoscience** with Princeton University historian of science Michael Gordin.*

4. How does Scott's description of the assumptions long inherent in techniques to search for exoplanets reveal the importance of rigorously examining our biases in the search for scientific knowledge?

III. Puzzles and Solutions

Solar system formation and shepherding moons

HB: You mentioned that there's not yet a complete theory of solar system formation, but what's our best understanding at the moment?

ST: Our best guess is that stars formed from collapsing gas clouds of interstellar gas, and that when they formed—because the gas that fell in had angular momentum—they were left with a spinning disc of material around them. That is probably pretty accurate, because we can see, although only at low resolution, similar discs around many young stars.

The next step would be to argue that, as the disc settles down, it starts to cool off: the star remains hot because of nuclear reactions, but the disc material around it begins to cool. Once it cools enough, below a thousand degrees or so, the heavy elements—things like iron and silicates—begin to condense out, because it's just not hot enough to keep them in a vapour-state.

And when they condense out, you get small grains of material that settle into the mid-plane of the disc—because that's the natural lowest energy state—where they're much denser than the gas, dense enough that they can begin to stick. So these small grains then collide and begin to stick together, and gradually build up into larger and larger bodies.

Now, starting with a grain that's maybe a micron across and going up to a planet like the earth is a pretty big stretch—that's many orders of magnitude in mass; and the details of how you get from the grains to the planets are still pretty obscure. We understand some of these ranges of many orders of magnitude, but there are others where you have to wonder how it's going to get from, say, a millimetre up to 10 kilometres.

So there are big gaps in our understanding, but roughly speaking, if you ignore those gaps, then you end up with these grains having condensed into solid bodies somewhere between the size of comets—10 kilometres across—to the size of earth or the asteroids.

Some of them then become massive enough that the gas around them becomes gravitationally unstable and they accumulate large atmospheres of gas; and then, eventually, as the star turns on, it becomes bright and powerful enough to blow the residual gas out. As a result, you're left with no gas disc anymore: just the star and these lumps that are left over.

HB: That's very comprehensive, thanks.

I'd like to talk about planetary rings and the effect that "shepherding moons" have on them. In particular, I'd like you to explain the specifics of that mechanism, but I'd also like you to talk more generally about the process of discovery, given that you were one of the people who played a seminal role in the development of our understanding there.

My experience is that a basic question many people have about scientists, particularly theoretical scientists is quite simply, *What do these guys actually do all day long?* Okay, they're studying the solar system, and they've got some equations somewhere based on some laws together with some data, but how do you connect these things? What do you look for? What triggers your imagination, and then how do you go ahead and verify that what you've done is actually correct?

ST: Well, the case of planetary rings is pretty atypical, so it's not a good thing to draw general conclusions from. I started working in the subject when I went to Caltech on a postdoc after I finished my PhD. I approached one of the professors there—Peter Goldreich—and said to him, "*I've been doing all this stuff on galaxies for my* *thesis, and I'd like to do something completely different. Is there anything you'd suggest?"*

And he said, *"Well, you could think about planetary rings."*

NASA's *Voyager 1* spacecraft was then on its way to Saturn and was expected to arrive three or four years later; and when it went past Saturn it was going to get a much better view of the rings than we'd had before.

HB: So, there was an opportunity for direct verification of some issues?

ST: There was hopefully an opportunity for direct verification, yes. But even more importantly, there was a puzzle. Of course, if there's a puzzle where things don't seem to be consistent with what you know, then that's a good sign that there might be some nugget of interesting science there.

The particular puzzle was that, if you look at Saturn's rings, you see a gap in the middle of them called the Cassini gap, because it was discovered by Giovanni Domenico Cassini in the 17th century, not long after Galileo first discovered the rings.

The reason that's a puzzle is that, if you have a ring made up of chunks of ice orbiting Saturn that, for some reason, never coalesced into a satellite, because the chunks of ice are continually colliding with each other (although at very slow speeds—a few millimetres per second), they're continually losing energy.

And if you have something in orbit that's losing energy but conserving its angular momentum, then it has to spread out in order to do that. And *that* means that any disc around a planet should always spread out—so if a gap would appear in the rings, the rings should spread out and fill it in.

So, the first question was, *Why is there a gap there?*

The clue, which had been recognized a long time ago, is that the gap was near a resonance with the innermost known satellite, Mimas, in the sense that material in the gap orbited Saturn in half the time that Mimas did. So the question was, *Is that a coincidence or is there something in the gravitational effect of Mimas that clears out a gap around a resonance like this?*

That was the puzzle, which we worked on for an extended period, eventually coming up with what we thought was a reasonable theory that could explain what was happening.

Basically, what was happening was that Mimas was transferring angular momentum to the resonant ring particles and pushing them out; and you could roughly explain the size and other properties of the gap with that model.

The relation to shepherd satellites is that, a couple of years later, before the *Voyager 1* spacecraft got to Saturn, a group of other researchers were observing an occultation of a star by Uranus.

Every now and then, one of the planets, in its orbit, passes in front of a star; and if you watch the star's light get blocked out by the planet, you can measure properties of the planet's atmosphere that are not accessible in any other way.

The trouble was that the occultation was only visible from somewhere in the South Pacific; and so they used a NASA aircraft with a telescope in it to fly along the correct track in the South Pacific to monitor the occultation. And what they discovered when they did that is that, about a few minutes before the star was scheduled to go behind the planet, the light from the star disappeared for a few seconds, came back up, then disappeared for another few seconds again and continued to do this for a few minutes.

Now, you can imagine how they felt when they saw their signal disappear for the first time. This is an event that only repeats once every few years, they were on this very expensive mission and they thought something had failed at a critical time. As it happens, they were recording everything; and there's a transcript where you can see what they said at the time, and it's pretty amusing.

At any rate, they were greatly reassured when they got the signal back. They got the occultation, and then the star came out, they saw the atmosphere on the way out, and then exactly the same pattern of dips occurred on the way out. Once *that* happened, it became pretty clear that it was a real signal, and that what was causing the dip in their signal was some sort of ring structure around Uranus.

But there was a problem, because these rings were typically only a few kilometres across. And if the problem of spreading that I mentioned with Saturn's rings was bad, it's *extremely* bad with Uranus. There's just no way that a ring that's that narrow shouldn't spread out and become diffuse in *much* less than the age of the planet. And what was particularly bad was that the rings had very sharp edges.

In any case, once that observation came about, we realized that if you just took the mechanism that we'd applied to Saturn and had a much smaller satellite much closer in, it could carry out this process of angular momentum transfer and confine a ring to a narrow region.

That came to be called "shepherding," because the satellite is like a sheep dog going around a flock of sheep, "barking" at it gravitationally to keep it all in line.

HB: But why do you feel that this was such an unusual process of discovery? Was it because of the combination of different levels of experimental verification? Because to me at least, in terms of the general recognition that there was a problem, trying to think of a mechanism that might be able to address that problem and then looking for ways of experimental verification of the model—that all seems, at least structurally, fairly standard.

ST: Well, it was unusual both because of the serendipitous discovery of rings at Uranus, but also because then, in 1980, the *Voyager 1* spacecraft *did* go past Saturn; and in the course of less than a week we learned more about Saturn and the ring system by orders of magnitude than we'd learned in the past 300 years.

The whole understanding of the ring system just improved so dramatically in a very short time—basically in the space of a week

—that we got direct verification of a number of features that we had predicted that was irrefutable and extremely clean.

And that's just simply not the typical experience in astrophysics, where you gradually develop a consensus on what nature is telling you through many false starts and over many years.

HB: It was a good choice of a problem.

ST: It was. Although he did actually warn me that it was a difficult problem, and it wasn't at all obvious that it wasn't a dead end. But at that point I was young and reckless, and it sounded like a really neat problem, so I was happy to try it.

HB: You began by saying that you had approached Peter Goldreich saying that you wanted something different to do from the galactic stuff that you had been working on for your thesis. Is that a characteristic feature of your approach, looking to switch domains?

ST: Not as much as some of my colleagues. I know colleagues who have worked on a subject area for 5 or 10 years and then just dropped it completely and moved into something completely different. Doing that is a challenge because you have a set of tools and intuitions that you've developed in one area that won't necessarily carry over to another area.

It typically means that there's a fallow period where you can't produce much as you get up to speed. It's also not something that's really encouraged by the current granting system, where once you get started you have to keep funds flowing to support your graduate students and postdocs. It's much harder to get a grant in a subject area where you don't have a track record than to continue doing things similar to what you'd done before.

Questions for Discussion:

1. Are you surprised at the notion that some scientists might reject working on a problem because it is considered "too difficult"? Might this be a contributing factor for why so many transformative discoveries are made by younger scientists? (Readers interested in this issue are referred to two other examples from Ideas Roadshow conversations where young scientists opted to pursue specific research ideas after having been explicitly warned of their inherent "riskiness" by their supervisor— Chapter 1 of **Indiana Steinhardt and the Quest for Quasicrystals** *with Princeton University physicist Paul Steinhardt and Chapter 3 of* **Cosmological Conundrums** *with University of Pennsylvania physicist Justin Khoury.)*

2. Why do you think the method of stellar occultation is such a common one in astronomical research?

3. Do you think that the scientific impact of unmanned probes such as the Voyager spacecraft are sufficiently appreciated by the general public? (Readers interested in an explicit comparison between the manned and unmanned space program are referred to Chapter 7 of the Ideas Roadshow conversation **Pushing the Boundaries** *with renowned mathematical physicist and polymath Freeman Dyson.)*

IV. Rings, Comets and Pluto

Mysteries, discoveries and evolving definitions

HB: You mentioned, when you were talking about Saturn's rings, that we're not exactly sure how these rings came to be to begin with. I'd like to explore that a little bit: what's our best guess as to why some planets actually do have rings and why some don't?

ST: That is a very difficult question. We don't really have any good idea. As you know, none of the inner planets have rings—neither Earth, Venus, Mars nor Mercury have them—whereas all of the outer planets have rings of some kind. But the rings around Saturn are far more spectacular, massive and luminous than those around any of the other giant planets.

Why Saturn and not the other ones? We don't know. *Why Saturn?* We don't know. There are various theories, such as: a satellite came too close to Saturn, was highly disrupted, and settled down to form a ring. There are theories that it was primordial—that is, that the rings formed at the same time as Saturn. We really don't know. There's also disturbing or worrying evidence that the rings are much younger than Saturn—maybe a few percent of the age of Saturn. Is that because their current configuration is unusual? We don't know.

HB: How do we know that, by the way? How do we know that the rings are much younger than Saturn?

ST: From theoretical arguments you can estimate how long the current configuration should last.

Because this mechanism for forming gaps involves transferring angular momentum, the shepherd satellite is transferring angular momentum to the ring, and we can calculate the rate at which that's occurring. We know the mass of the satellite, so we know how fast it should be receding from the ring—and the time scales for that recession, if we extrapolate backwards, put the satellite in the ring at times that are only a few percent of the age of Saturn.

That may be misleading, however. For instance, it could be that the generic process for the ring is that angular momentum is transferred out from the ring to the satellite, and as the ring particles get further and further out, they become more gravitationally unstable and then they condense into a new satellite, and then that gets pushed out. But nobody has been able to make that theory work quantitatively yet; and, again, with only one example, it's a little hard to know what's an accident and what's not.

Perhaps the optimistic way to think about it is that Saturn's rings represent a very unusual and rare occurrence, and in the *Michelin Green Guide to the Galaxy*, that's the only reason that we've got three stars.

HB: You're a very careful scientist; and I know you don't generally like to indulge in speculations—at least not without many caveats specifying that you're indulging in all sorts of speculations—but I'm going to ask you to indulge in some anyway. Do you have any intuition as to what might have led Saturn to be as different as it is? Do you have any pet theory or pet ideas?

ST: No, I don't.

Again, it's very hard when you've only got one example, and there's nothing obvious about Saturn that's so different from the other giant planets. My personal guess is that it's some sort of unusual and rare event that could have happened to any of the giant planets but Saturn's the only one that it happened to. As to the nature of that event, or exactly how lucky we are to have a ring system like Saturn's, I have no idea.

HB: Do we have any current ability, or can you foresee any ability in the longer term, to get a sense of possible ring structures of exoplanets?

ST: That's possible, and people have certainly thought about that. The typical problem is that the exoplanets we can see best are the ones that are transiting. Those that are transiting can only be seen because they pass in front of the star, and they only pass in front of the star if their orbit is edge-on to us. So in the likely case where the planet's equator is aligned with its orbit, any rings would be edge-on to us, in which case, we won't be able to see them. I think it's quite possible that many of the planets we have seen have beautiful ring systems, but they're all edge-on and we can't see any of them.

HB: Is there any future technological or higher-level, diagnostic technique that we might be able to imagine which could help us there?

ST: You could imagine that, if you could separate the light from the planet from the light from the star, something like Saturn would appear bigger than it should because you're seeing the light from the ring system. The spectrum of the material in the rings— because it's solid ice—would look different from the spectrum of the gas in the planetary atmosphere. Maybe something like that, then, in principle, might be tried.

The most interesting recent ring observation, which was also a surprise, is that there is an asteroid orbiting in the outer solar system, which occulted a star a couple of years ago visible from South America, which has a pair of rings around it. So, somehow, the asteroid, like Uranus, was able to form narrow rings—and, again, we have no idea why that is.

HB: How big is this asteroid?

ST: It's maybe 100 or 200 kilometres across, so it's a very different-sized system from Saturn or Uranus.

HB: I'd like to move now to comets—another area where you've done seminal work. Tell me what we know about comets, where they come from, and what sort of mysteries are associated with them, as well as your particular experiences related to the process of discovery.

ST: Well, everybody knows what a comet is, roughly speaking: it's this beautiful thing that has a nice, long tail in the sky. The way to think about them from a theoretical point of view is as an intermediate stage of the planet-formation process I was talking about earlier.

Recall that I was talking about this process that started from dust grains, accumulating all the way up to planetary scales. One of the intermediate stages is bodies a few kilometres or tens of kilometres or hundreds of kilometres across. Some of those bodies will be accumulated into planets, but not all of them, because the process isn't all that efficient.

So the right way to think of an intermediate stage of the formation of our solar system, or other planetary systems, is as a series of a small number of planets with huge numbers of small bodies left over in between. Those small bodies are then subject to the gravitational influence of the planets, which leads their orbits to evolve chaotically—and in some cases those bodies will be ejected onto very eccentric orbits going far outside the planetary system. Normally, you'd say that those orbits couldn't survive, because every time they come back again, they get another kick from a planet and eventually it'll kick them into escape energy, and they'll be lost to interstellar space.

But in the case of the solar system, with our existing configuration of giant planets, as the orbits get large enough, external influences—gravitational kicks from passing stars, the gravitational field of the overall galaxy—begin to change the orbits, so that they no longer come within the planetary system.

And once they don't come close to the giant planets, they don't get any gravitational kicks from the giant planets, and their orbits are kind of frozen in place. They can stay frozen in place for billions of years, and then some other passing star or some other effect of the galactic, gravitational field will cause them to come back into the planetary system, and then they'll appear more or less out of nowhere and come close to the sun.

They've got a mix of frozen ice and rock; and as they get close enough to the sun, the ice starts to sublimate, and that's what creates the tail. So, there's a steady flow of new comets that we get from these random effects that we can't really calculate. Once the comet starts losing material through heating from the sun, it will typically last a few tens or hundreds of orbits and then disintegrate or become inactive—all the frozen ice and other gases are sublimated so it doesn't produce a tail anymore and it's very hard to see—and then that's the end of the comet.

HB: Once again, I'd like to detail how we've figured this all out. How do you actually sit down and develop, based upon what we understand at the moment, a prediction that these things are going to be coming from this particular place and they're going to be interacting in this way, as opposed to that way? Talk me through a little bit of your process of discovery with respect to the Kuiper belt, and all of that.

ST: I think the best example of this is not my process of discovery but a somewhat earlier process.

The origin of comets was a mystery for a long time, and the seminal work on this was done by the Dutch astronomer Jan Oort in 1950. It's possible to make very accurate measurements of the orbits of comets; and people looked at these orbits and they seemed to come from all over the place with no clear signal, no clear way to interpret where, exactly, they were coming from.

Oort recognized that, since the comets were coming from a long way away, the important thing was not the orbit that they had when we saw them, but the orbit that they had *before* they entered the planetary system, which is different because of the gravitational kicks induced by Jupiter and Saturn.

Oort and his collaborators followed the orbits back in time until before they had entered the planetary system. This was not easy, of course, because there were no digital computers back then, but they had techniques for doing that. And when they followed the orbits back, they discovered that all of the orbits, as measured from outside the solar system, came from a distance from the sun of about 20,000 to 50,000 times the distance of the earth from the sun.

That's a very long way away, and the comets couldn't have possibly formed there because the coalescence of grains to make solid bodies only occurs in high-density regions. Out there, even if there was a disc or other material left over from the formation of the star, it would be far too low-density to produce anything, which means that the comets had to have come from within the planetary system and somehow gotten kicked out to that volume, which is now called the Oort cloud. Once you had *that* clue, the step to say that the kicks must have come from the giant planets is not really such a difficult conclusion to arrive at.

The particular work I was doing on what's now called the Kuiper belt arose because we were trying to do similar things. Given Oort's model, and given that computers were constantly improving, it was a fairly straightforward idea to say, *Let's take a bunch of comets in the Oort cloud, let them come in, follow them as their orbits evolve due to the influence of the planets, and see if the distribution of orbits looks like what you see.*

That was hardly "rocket science", as they say, but when we did that, we found that there was a mismatch: the comets that we produced from the Oort cloud were, roughly speaking, spherically distributed, but a large fraction of the comets that we see have orbits that are very close to the same plane as the planetary system.

If you think about it, it's very hard to concentrate the orbits that are initially spherical down to the plane. Once again, you didn't have to be a rocket scientist to figure out that there had to be some source for these other comets that wasn't the Oort cloud, and was probably directly associated with the planetary system—because it was flat, with the same orientation as the planetary system.

Once you think about *that*, the only logical way to do this is to have a belt of cometary material outside the planets, far enough away that it wouldn't have been detected, but close enough so that the gravitational kicks from the planets could gradually cause comets to diffuse out of that belt into the inner part of the planetary system.

HB: And my understanding is that Pluto is one of these guys that's in that region, right?

ST: Well, of course, you can call Pluto whatever you want to call it.

HB: Well, that's really what I'm getting at.

ST: So, there are lots of comets in the Kuiper belt that have orbits just like Pluto's, a composition just like Pluto's, all the other properties just like Pluto's; the only difference being that Pluto is bigger than most of them, but not all. So you can call Pluto a planet, but it makes a lot more sense from any standpoint, to think of it as one of the largest members of this belt of objects.

HB: Oh. I didn't realize that. There are some that are just as large as Pluto?

ST: There are some that are larger than Pluto. It's a little hard to be sure, because they're more distant and it's harder to resolve the size, but the best evidence is that Pluto is *one* of the largest members, but there are a handful that have been discovered already that are bigger than Pluto.

HB: So that's the primary motivation for this "downgrading" of Pluto that happened 10 years or so ago, presumably.

ST: Well, the obvious problem that people who do this sort of thing were worried about is that, if something is bigger than Pluto, you can't really call Pluto a planet and not call that other object a planet as well. Then, if you discover more of them, where do you draw the line?

HB: Was the public portrayal of the situation at all frustrating for you? My recollection is that the way the whole Pluto thing was framed was somewhat arbitrary: *One day astronomers said Pluto was a planet, while the next day it somehow didn't measure up, it wasn't good enough.*

Whereas it seems like this would have been a good opportunity to really explore *why* this has happened and tie it into our understanding of this development in astrophysics and astronomy. Did that actually happen, am I misremembering? Or was it more or less as I remember?

ST: I think you're correct: it was what you would call a "teachable moment"—any time your conventional definitions fail, any time that you've got something that doesn't seem to fit into your conventional definitions, you learn something very important.

Breakdowns of conventional definitions in science are thus opportunities rather than problems. I think that definitely was lost in the media circus about Pluto.

I remember there was a *New York Times* headline that said, *New York's a Tough Town If You're a Planet*, because a lot of the initial impetus for change came from the opening of the Hayden Planetarium, when Pluto wasn't counted as a planet.

So, yes, I thought it played out a lot less well than it could have. My attitude, and I think the attitude of most of my colleagues, was that you can call it anything you want to, but the important thing is the properties. It would have been nice if the media attention attached to this had focused more on the positive view that you really learn something when your old definitions break down.

Questions for Discussion:

1. Are you surprised that precise calculations of orbiting comets were able to be done in the days before digital computers? What sort of techniques do you imagine existed to enable such calculations to occur?

2. Do you think that we will one day be able to detect rings around exoplanets?

3. Do you remember hearing about Pluto being "downgraded" as a planet? Did you have a clear sense of why that occurred? Do you agree with Scott that it represented a missed opportunity for "a teachable moment"?

4. How would you describe Pluto now after having read this chapter and why it is no longer considered a planet by most astronomers?

V. Investigating Stability

Considering past and future

HB: Two final questions before we leave the planetary-scale realm. You referred briefly to this point earlier, but I'd like to highlight it for a moment. Our solar system is composed of four solid planets—Mercury, Venus, Earth and Mars—followed by four gas giants—Jupiter, Saturn, Uranus and Neptune. That seems a bit of an odd formation, when you think about it. Of course, we have asteroids of rock that are way out there, much further than any of the gas giants, but it seems like a rather odd structure to have a whole bunch of gas giants that are lined up close to the end. Is there some theme there or reason for that?

ST: It's certainly possible to have gas giants much closer in. The existing studies have found gas giants with orbital periods as short as a few days around the host star—hundreds of times closer to the host star than our gas giants are.

Is our system all that different or unusual? Well, we don't know. As I said earlier, we only could barely begin to detect our solar system with current technology if we were looking from outside the sun. Right now, we're finding planets around 10% or 20% of stars, depending on how you define it. It could be that the other 80% to 90% look just like our solar system. If that were the case, we wouldn't be unusual at all.

However, it could be that *nothing* looks like the solar system, and we're *incredibly* unusual. I don't think that's likely because, although we can just barely see Jupiter, we can see planetary systems with planets like Jupiter somewhat closer, and there seem to be plenty of those.

One possible explanation as to why we have this particular configuration is that it's very hard to form a planet like Jupiter very close to the host star, so it's most likely that the ones we see close to the host star formed at large radii and then migrated inwards due to interactions with other planets, interactions with the star, or interactions with the protoplanetary disc. We don't know which of those is right, but if a Jupiter-like planet did migrate inward, you could imagine that it would be pretty unhealthy for any terrestrial planets like Earth and Mars that were in its way.

So, it's possible that only a small fraction of planetary systems avoid migration of Jupiter-like planets inwards, but that those are the ones that maybe are most likely to have life, because the planets in the habitable zone are likely to have been destroyed via collision with the gas-giant planet as it migrated inwards.

HB: That brings me to a more general question about the stability of the solar system. I could imagine that if I have a whole bunch of bodies in motion it would *never* necessarily coalesce into any sort of stable form, which our solar system seems to have. Why is it the case that our solar system seems to be so stable?

ST: Well, that's a good question, and a very old one. Newton certainly thought about it. He understood that, in principle, he could calculate the behaviour of all the planets in the solar system based on his law of gravity and his law of motion, but the only system that he was able to solve exactly was a simplified version in which there was just the sun and one planet: the two-body problem.

In the case of the n-body problem, he was aware that this was determined by his equations, but he couldn't solve them. At that point, there are basically two options.

One is to say, as you indicated, that the gravitational effects of all the planets on one another will gradually build up irregularities in the orbit's eccentricities or inclinations until eventually the planets collide, or one of them falls into the sun, or it gets ejected into interstellar space.

The other possibility is that the gravitational influence of one planet on another is periodic, so they induce periodic oscillations in the orbits but the oscillations never grow.

In the state of the understanding of physics at the time of Newton, we didn't know which of those options was going to be correct. Newton's view was actually quite different. As you know, he had rather complicated theological views: he believed that God not only created the universe, but actively intervened to make it do whatever He wanted. His view was effectively that God had a "service contract" with the solar system, that the motions would gradually get more and more irregular before God would step in and do a tune-up and fix it.

On the other hand, Leibniz, who was a contemporary of Newton's and had controversies with Newton about other issues—such as the invention of calculus—believed that the system was stable. He famously said, "*Professor Newton's views are very strange. He believes that if God made a watch, He somehow wouldn't have the foresight to have made it good enough so that He wouldn't have to wind it from time to time.*"

Now, of course, we've moved beyond that sort of dialogue, but this has been a problem which has attracted the attention of a lot of very strong mathematicians and physicists who have proved lots of theorems. This gave rise to the whole modern field of nonlinear dynamics; and although we understand things qualitatively much better than we used to, the actual question of whether the system is stable can only be decided by running a numerical calculation.

With modern computers you can do that—to follow a solar system throughout its whole life would take a few weeks on your laptop—and what you find when you do that is that, first of all, the system is chaotic: if you change the initial conditions slightly, the two solar systems will diverge exponentially—it's very sensitive to the initial conditions.

For example, the fact that you came here today changed the tidal field from your mass on Jupiter enough so that, in 7 billion years, when the sun turns into a red giant, there will be a corresponding uncertainty in the position of Jupiter's orbit.

HB: This is "the butterfly effect".

ST. That's right: it's the butterfly effect. So that's certainly there.

On the other hand, when you do the numerical calculations, in about 99% of the solar systems that you try that are consistent with the errors in the known initial conditions and parameters, everything is fine until the sun turns into a red giant and incinerates the equivalents of Venus, Mercury and maybe even the earth, at which point there's not much point in trying to follow things past then.

In the other 1%, Mercury's orbit would get sufficiently irregular and would either collide with Venus or the earth or fall into the sun, so the empirical answer seems to be that there's a 99% chance that the system is stable.

HB: 99% is a pretty conspicuously large value, so I'm wondering if there is some structure somewhere that might enable us to somehow see that without actually grinding through the calculations.

ST: It's a good question. The only way we can tell right now is to grind through the calculations. The reason to be pessimistic about finding a simple structure or a formula that would enable you to predict this without doing all the work is that the result seems to be quite sensitive to the detailed configurations.

People have tried experiments where they do it with general relativity or without general relativity, or they change the masses of one of the planets by 10%, and the results can change dramatically. So there may be a structure but, if so, it's something that's very sensitive to the detailed configuration of the solar system.

The other thing to bear in mind is that we're talking about the system being stable when you calculate its future behaviour, but that's no guarantee that it was stable in the past.

One natural way to think about it, for which we have no direct evidence, is that, when the system was maybe 10% of its current age, it had more planets and they had shorter lifetimes, so they already collided with the sun or one of the other planets or got ejected, in which case we wouldn't know that it had happened.

This is what some physicists call "self-organized criticality": you have a system that's stable on, say, a 100-million-year time scale and readjusts itself by getting rid of some planets so it's stable on a billion-year time scale; then it readjusts itself again by getting rid of one or two more so it's stable on a 10-billion-year time scale and that's as far as we've gone.

If the sun didn't turn into a red giant, presumably what would happen is you would get rid of Mercury in 20 or 30 billion years, and you'd have a more stable system. If you then went for a 1000 billion years, you might get rid of something else: it's just extremely slowly settling down to a more and more stable state, until eventually only one planet is left.

Questions for Discussion:

1. Might there be some ways of indirectly detecting that there used to be more planets in our solar system?

2. What do you think Howard has in mind when he talks about "some structure somewhere that might enable us to somehow see that without actually grinding through the calculations". Are there examples of similar sorts of "structural awareness" occurring in the history of science?

3. What sorts of assumptions about the nature of life-supporting biology is Scott making when he talks about "the habitable zone" for planets? Do you think that those assumptions are warranted?

VI. Large-scale Issues

Colliding galaxies and dark matter

HB: I'd like to turn to galaxies now. Let's start with a basic question. I understand that galaxies come in a few different flavours: galaxies that look like discs, galaxies that look like spirals and one other type that suddenly I can't remember...

ST: Discs, spirals and elliptical galaxies.

HB: Right. So, why is that?

ST: Roughly speaking, if you look at galaxies, there are two flavours.

There are galaxies that appear to be like a spherical, or elliptical, or ellipsoidal ball of billions of stars—structures that are basically held together by the self-gravity of this huge assembly of stars, and each star is orbiting in the combined gravity of all the other ones.

There are also disc galaxies—some with spirals, some without—in which, instead of this spherical structure, there's a more or less flat disc that contains stars, gas, ionized gas and a variety of other components.

Why is there a difference between them? We don't know.

But one hypothesis is that if you have two spiral galaxies that come close enough together, the two will merge, and this merging process will be quite violent: it will churn up the stars so that the flat discs will get churned up into a roughly ellipsoidal structure, the gas will lose its angular momentum—because of the irregularly, violently varying gravitational field—and fall into the centre, where there'll be a burst of star formation; the winds and supernovae from the new stars will blow out all the remaining gas, and then those stars will settle down to be part of the elliptical galaxy.

That model probably has a great deal of truth in it. It doesn't necessarily explain everything, but it gives a rough idea of why there might be two flavours.

HB: This model seems to imply that these collisions between galaxies happen with some regularity, is that correct?

ST: Yes, that's also a good question. It's not at all obvious that that should happen; and in the early studies of galaxies where people thought of these as analogous to particles in a gas, randomly moving around through the universe, the collision rate was extremely low.

The reason that's no longer considered to be true is that in our modern picture the galaxy is basically gravitationally unstable—that is, the velocities of these galaxies relative to the overall cosmic expansion are quite low, and galaxies grow by a kind of hierarchal, merging process: you make a small structure with a few stars, there's another structure somewhere else, they become gravitationally attracted to one another and they fall together and merge.

So, rather than being a rare event, this type of "merger" is now considered to be the basic building block in making the galaxies to begin with.

HB: So, let's talk about cosmology now and I'd like you to go all the way back in time and give me a sense of how these galaxies, or at least first chunks of galaxies, formed to begin with.

ST: Well, in the standard, Big Bang cosmology—which, in most respects, is now extremely well verified by the data—everything starts off very hot, and as the universe expands, the material cools.

The first, critical step is when the protons and the electrons —which comprise most of the baryonic material in the universe —recombine to form neutral, hydrogen atoms. At that point, the hydrogen is decoupled from the gas of radiation, and so it can start to condense.

Initially, then, you have a homogeneous universe full of this relatively cold hydrogen and you start to form gas clumps. The collapse is enhanced by the fact that there's also a background of dark matter, which contains most of the mass, so the dark matter and the gas collapse together into small lumps. The gas can cool, and so the gas shrinks much more than the dark matter material; and then these lumps start merging together. As the universe continues to evolve, larger and larger lumps coalesce together and the gas in the centres of these lumps eventually form stars, and from there you have galaxies. But the star formation typically occurs long after the dark matter and the gaseous material have already coalesced as a separate structure.

HB: Tell me a little bit more about that star formation process and how it actually works; or what our best guess at understanding that is.

ST: Again, that's one issue where we don't understand things very well. If you look at all the systems that we know of in astronomy— stars, planets, galaxies, the universe—paradoxically, the one we think we understand the best is the universe, because it's so hot that the physics is a lot cleaner.

The star formation process we believe is occurring because the gas is cooling; as it cools, it loses energy; as it loses energy it can't support itself against its own gravity, and so the material begins to condense.

That's a rough view of how the star formation occurs. It's basically because—in contrast to the dark matter—ordinary matter, like hydrogen, can cool—and as it cools it becomes denser and denser under its own self-gravity.

HB: You've mentioned dark matter on several occasions, and if I'm a non-specialist listening to this, I'm hearing Professor Tremaine talk about dark matter as if it's out there, a contributing factor that seems just as real as everything else. Tell me why you're confident that that's the case—or at least why the people who are invoking the models that you're describing have such confidence in its existence.

ST: Well, it's certainly true that not all of my colleagues are totally confident that dark matter exists; and although there's a huge amount of indirect evidence for it, there's no direct evidence—that is, we don't know what it is; and until we know what it is, I don't think anyone will really be 100% confident that it exists. But the indirect evidence is very strong.

Part of the evidence is that we can measure the distribution of mass in galaxies and it simply doesn't correspond to the distribution of the mass that we can see. We know what the masses of stars are, we can make a pretty good estimate of the total contribution of the mass due to stars in a galaxy. If we look at orbits of satellite galaxies, if we look at the rotation speed of hydrogen gas in the galaxy, if we make direct measurements by looking at distortion of light from background galaxies by the gravitational field of the galaxies—all of those give an extremely consistent picture, which says that there's up to a factor of 10 to 30 more mass in the galaxy than you can ascribe to the stars and to the gas.

You can hide that in other forms—such as black holes and other putative structures, such as so-called "bricks" of matter—but first of all, almost all of those forms have been ruled out by indirect arguments; and secondly, there's cosmological evidence from the primordial distributions of heavy elements formed early in the universe that the density of anything that can contribute to the synthesis of heavy elements is much lower than the mass that we see in galaxies.

So those are two indirect arguments for dark matter. There's another one as well. You recall I said a few moments ago that early in the history of the universe the hydrogen was ionized, composed of individual protons and electrons. Those are strongly coupled to the radiation, and that means that any initial fluctuations in the density distribution of the universe would be homogenized by a process called "Silk damping," after the physicist Joe Silk, which basically means that the coupled radiation and hydrogen fluid is dissipative and it tends to get smoothed out.

The dark matter, however, is *not* coupled to the radiation—that's why it's dark—so its fluctuations in density wouldn't be smoothed out.

So unless there was dark matter, by the time the hydrogen became neutral, it would be so smooth that you wouldn't have been able to form any galaxies in the time available.

So there's a set of several different lines of argument, all of which are solved very easily by postulating the existence of dark matter, and each one of which is very hard to solve without that.

Having said that, many searches have gone on for dark matter in many different forms. The most popular view now is that it's some new elementary particle that is not part of the standard model of particle physics that we hope to detect by very sensitive observations.

But it's also true that people have been looking for this for many years, the current limits on the cross-section on the interaction strength of that particle have gone down by many orders of magnitude, and they still haven't found it. Obviously it's correct to keep looking, but at some point, we may have to say, "*If we can't find it, what other alternatives are there?*"

HB: Do you have any other alternatives in mind right now? Do you have anything that you're willing to speculate on?

ST: Well, some people have tried to speculate that there is some failure in Einstein's theory of general relativity at the largest scales.

But it's very difficult to tinker with the theory of relativity, as you've probably discovered too. It really is a kind of unique theory, and it's very hard to add any bells and whistles to it that don't either create a fundamental inconsistency in the theory or that are ruled out by the extremely accurate tests that show that general relativity is correct on the scale of the solar system.

So, one possibility is to modify gravity, but I think that's not very attractive. The other possibility is that we simply haven't been imaginative enough in thinking about alternative, elementary particles. The third possibility is that we *have* been imaginative enough, but we don't really know the interaction strengths, and if we keep going for another order of magnitude or two, we'll find it.

Questions for Discussion:

1. What, precisely, is Scott referring to when he talks about attempts to modify Einstein's general theory of relativity? Readers interested in a deeper analysis of this issue are referred to Chapter 6-9 of the Ideas Roadshow conversation **Cosmological Conundrums** *with University of Pennsylvania cosmologist Justin Khoury.*

2. Do you think that we will have "direct evidence" of dark matter in the next 5 years? If not, is there a time at which a lack of direct evidence for dark matter can serve as a form of evidence **against** *the theory?*

VII. Black Holes

Different types, different evidence and open questions

HB: I'd like to talk now about black holes—specifically, these supermassive black holes that are now thought to be at the centres of galaxies. Of course, there's a fairly long and interesting history of the idea of black holes, even from a Newtonian perspective, but it's probably worth stating that for many years they were thought to be somewhat speculative objects arising from the general theory of relativity. For decades, there was some real debate as to whether there were any actual black holes out there, then they started to find candidates and confidence began to steadily build that such things actually existed.

Now, we move a few decades into the future, all of a sudden, not only are black holes everywhere, but we're told that there are these huge, supermassive black holes that are at the centre of many—if not all—galaxies.

So, putting myself in the position of the curious non-specialist, my first question is: *Why shouldn't I be sceptical about all of this?* After all, at first, most scientists seem convinced that no black holes existed; and now, a few decades later, there are apparently these mammoth ones all over the place, in the centres of every galaxy. Why are you guys suddenly so confident?

ST: Well, of course, you *should* be sceptical; having an educated public that's sceptical about claims that scientists make is very important.

Let me start by making a couple of points. First, although black holes do have a long history, and they're an inevitable prediction of Einstein's theory of relativity, they were only really understood 50 years after the theory was put forward. In fact, I'm told that Einstein never really believed that they exist.

Formally, they're a singularity in space-time, which, it turns out, is always surrounded by an event horizon—a one-way membrane—so that things can go into the event horizon but they can't come out.

All of those properties, I think, were pretty well understood 30 or 40 years ago. As you say, they were discovered before there was very strong evidence for astrophysical black holes, but I think it was clear from the start that you can't make a black hole in the laboratory.

So, if you're going to study the properties of black holes and the properties of Einstein's theory in strong gravitational fields, you have to look to astrophysics to see if nature has been kind enough to provide some black holes that you can measure.

The second thing I'd say is that the situation is somewhat analogous to what you described for dark matter: nobody has seen a black hole, there's no direct evidence for a black hole. That's not surprising because they're black, and they're also very small; but the evidence that black holes *do* exist is so strong that I think almost all physicists and astrophysicists believe that they exist and that we can point to locations where they're found.

As you said, there are two flavours: there are stellar-mass black holes created in the collapse of massive stars, but there are also these so-called supermassive black holes, which are found in the centres of galaxies.

The arguments for those basically arose from the discovery of quasars, so let me say a little bit about quasars first.

If you look at galaxies, most of the light comes from stars not that different from the sun. But in a small fraction of galaxies, there's a point-like source of light exactly at the centre of the galaxy. You can tell from the spectrum of that light that it's coming from hot gas, but not from stars. The origin of that light was a mystery for many decades.

The most extreme examples of that are quasars. They're called quasars, because it's a contraction of "quasi-stellar source," and they're called "quasi-stellar" because they were originally found as things that looked like stars but the spectrum didn't look like a stellar spectrum. The key advance came in the early 60s. Some quasars are also strong radio emitters, and if you get a radio source like that, the first thing you do is try to find it with optical light, because the optical telescopes have much higher resolution and you can get a lot more information.

But radio telescopes don't give you very good positions, so they knew roughly some region where the quasar was, but they couldn't figure out which star it was. That problem was solved because it happened that there was a particular, bright quasar that was in the path of the moon; and at one point—in 1963 I believe it was—they realized that the moon was going to cross in front of this quasar.

Since we know the moon's orbit very accurately, if you can measure the time at which the light from the quasar disappears, you know exactly where the quasar is—you have an accurate position. Once you had the accurate position, which turned out to have been measured in Australia, you could identify which star it was—and the optical astronomers at Caltech were able to get a spectrum of it.

The person who had the spectrum, Maarten Schmidt, puzzled over it for a long time and finally realized that it was an absolutely standard hydrogen spectrum, but it was red-shifted—meaning that it was displaced from the usual wavelengths of the hydrogen line, which implied that it was travelling away from us. It turned out that this object was travelling at, I believe, 14% of the speed of light, which, in turn, meant that it must be very far away and participating in the cosmological expansion.

So, it looked just like an ordinary star, but it had to be something like 100 billion times brighter. Once they recognized the extreme luminosity of the quasars, it was clear that they were in galaxies, but they just hadn't been seeing the galaxy because it was lost in the glare of the quasar light. Then they had to try and figure out how something can shine so brightly; and the arguments that it's a black hole start to multiply.

First: black holes are much more efficient at converting mass into energy—probably a factor of 100 times more efficient than even a nuclear reactor—and if you don't have a black hole producing that luminosity, that rate of energy emission for the lifetime of a galaxy, it's not clear where it could come from, because it requires more fuel than even a galaxy can provide.

Second: the quasars vary rapidly, and that means they have to be very small, because otherwise you can't coordinate the variation across the size of the source—and the only objects we know of that can produce that much energy in that small a volume are black holes.

Third: in X-rays, they can measure spectral lines of some elements; the lines are grossly distorted and the only way they can be made that broad and distorted is if the gas is emitting from deep in the bottom of a gravitational well.

The arguments can go on, but enough of them are now present, and all the alternative models have become so improbable that it's absolutely clear that, by far, the best explanation for quasar power is black holes.

Once you know that, you can calculate how many black holes should be around that used to be shining as quasars but maybe have lost their fuel supply, or for one reason or another are now dead, and then you can look for those in nearby galaxies.

Those arguments motivated people to go look in nearby galaxies for massive concentrations of material at the centre. That was, in fact, one of the original, primary goals of the Hubble Telescope when it was launched.

By now, with the Hubble and other instruments, people have looked in the centres of perhaps close to 100 galaxies; and in most cases, they find large concentrations of dark mass, which almost certainly is a black hole.

In the best cases—the best case, in particular, being our own galaxy—you have maybe 4 million times the mass of the sun in a volume that's not much bigger than the size of the solar system, and the only thing we know of that could be providing that is a black hole.

HB: That's a great summary, thank you. I have three questions. My first question is, once again, on behalf of the sceptical, non-specialist. I can imagine that somebody who's not an expert on this might confuse the darkness from a black hole with dark matter, so I think it's probably worth pointing out explicitly that we're talking about something completely different here.

ST: Yes. If you look at most of the volume of a galaxy, most of the mass that you see is comprised of ordinary stars. If you look at very large distances outside the stellar system, there is extra mass, which is what I'm referring to as "dark matter." If you look at the very centre of the galaxy, far inside the distribution of most of the stars, there is also extra mass, which is a black hole. Now, you might wonder why there are two categories and why we don't just refer to them as the same thing but it's very clear they're different.

First of all, the mass in the black hole is very concentrated and it's only a fraction of a percent of the mass of stars whereas the dark matter, in the outer parts, is 10–30 times the total mass of the star.

Just in terms of total mass, then, they're very different; in terms of total density, they're very different; in all of their properties, they're really quite different. The only property they share is that they're both dark—but that's not a surprise because, if they were bright, we would have found them a long time ago.

HB: Very good. Next question: granted that, as you've explained, we have all sorts of evidence that points to the fact that we have those huge black holes in the middle of galaxies, but I would then like to have a theory, which says that these things naturally and inevitably—or, at least, statistically—form in that particular way. Do we have such a theory?

ST: No, although I think not all of my colleagues would agree with me. I think the current state of the theory is really to ask, *If you have a dense, stellar system, then what can it do?*

So, hypothetically, you would make a dense, stellar system, like you might see in centres of these objects, and then you would ask yourself how it would evolve. Well, some of the stars will collide and merge, forming bigger stars; and those stars will then tend to collapse and form small black holes. If you have stars or small black holes, they'll lose energy by gravitational radiation, get smaller and smaller orbits and then coalesce—they'll form even bigger stars, and even bigger black holes.

You can go through a whole set of these arguments and the general conclusion is that whatever you do through a wide variety of possible evolutions, it eventually ends up as a black hole. I don't think we have a persuasive theory as to why there are black holes there, but every other system that we can think of seems to eventually end up as a black hole.

What some of my colleagues would say is that the issue is not finding a theory that convinces you that you have to make a black hole, it's that black holes are such a natural outcome of many different evolutionary paths that the real challenge would be to find something that *doesn't* produce a black hole.

On the one hand, people might say that black holes are so weird and that extraordinary claims require extraordinary evidence, so you have to have really strong evidence that it's a black hole, but given the experience with the robustness of black holes in general relativity and with the necessity of needing black holes to power quasars, I think it would be fair to make the argument that it's not an extraordinary claim at all: it's exactly what you would have expected.

The onus is then on people who think that they're not black holes to prove that there's some other, viable alternative.

HB: Alright, so let me be very simplistic for a moment. Again, without knowing anything about general relativity or anything like that, I know that gravity attracts, I've got a whole bunch of stuff that's a dense collection of stars and all the rest of that and that, over time, it would start getting denser and denser, particularly in the core, and I'd start getting to these things that you're calling a black hole.

Now, if I wanted to test that theory—which is to say, if the things that we are finding empirically are roughly the size of that black hole that you would imagine, I should be able to back-date it to our understanding of cosmology, how long these things have been around and what size we would imagine your average black holes would be at this particular time.

So I would imagine there being some ability to verify, not just that it's a black hole, but also that it should be, roughly, after this amount of time, statistically, a black hole of this or that kind of size.

In other words, the basic idea that you've got a bunch of stuff and you have a black hole in the middle because that's the end-state when you have a bunch of stuff that's attracting doesn't seem sufficient to me. I could imagine an entire galaxy that's one big black hole after a certain amount of time; or I could imagine, after a certain amount of time of evolution, a much smaller black hole.

You would think that there should be some ways of saying, *"OK, given how old the universe is, given how old, statistically, from when galaxies have been forming and so forth, you should be able to put those things together and not only realize that there's a black hole in the middle but it's roughly the same sort of size as one might expect from cosmological arguments."* Is that right? Or is there simply somehow too much theoretical latitude to be able to say anything explicit there?

ST: Well, you've hit on, I think, a couple of the really active, unresolved questions in this whole process.

The first issue, which I think is somewhat worrying, is that we now see very large black holes—billion solar mass black holes—at redshift 7, when the universe was less than a billion years old. And when you try to construct those models, it's typically rather difficult to build big black holes so early. What that means is that, somehow, galaxies seem to be able to build black holes on a timescale that's less than 10% of the cosmological timescale. We don't understand how they do that but, on the other hand I'm not so worried because, as I said earlier, we don't know how to form stars, we don't know how to form planets, we don't know how to form galaxies—so I'm not so worried if we don't know how to form those particular black holes.

Secondly, what it seems to me that you're basically asking is, *"Why are the black holes a few tenths of a percent of the mass of the galaxy rather than much bigger or much smaller?"*

HB: Exactly.

ST: That's a very fundamental question, which I think is probably *the* fundamental question in galaxy formation, but I'd phrase it a little differently: *"Does the galaxy determine the properties of the black hole, or does the black hole determine the properties of the galaxy?"*

The black hole is only a tiny fraction of the mass of the galaxy—as I said, a couple tenths of a percent—but because a black hole is such a condensed object, the energy that was released in forming the black hole is much larger than the energy that was released in forming all the rest of the galaxy.

For example, if you make a typical black hole in the centre of a galaxy and you feed back 1% of the energy that was released in forming the black hole and couple that to the gas in the galaxy efficiently, you can blow all the gas in the galaxy out. And if you did that, there would be no gas left and the black hole would stop growing.

So the problem for understanding galaxy formation and black hole formation is that a tiny, inefficient coupling at the level of

1% between the black hole formation and the gas in the rest of the galaxy can completely change the evolution of the rest of the galaxy and the black hole.

We see indirect evidence for this happening: many black holes have jets of material that are going out into the interstellar medium, and there's evidence for bubbles around the ends of these jets which suggests that there is hot gas that's been heated up by the impact of the jet from the black hole.

So, until we can understand the feedback at the 1% level, we're going to have a very hard time understanding, in detail, the relation between the evolution of the black hole and the evolution of the galaxy.

Questions for Discussion:

1. How can be certain that an object that is travelling so quickly (like the quasar Scott mentioned that was determined to be moving at 14% of the speed of light) is so far away from us?

2. Might it be possible that some galaxies do **not** contain a "supermassive" black hole at their centre? If so, what might that imply for better understanding the relationship between supermassive black holes and their surrounding galaxies?

3. While it is clearly unrealistic to "make black holes in a laboratory", to what extent might we be able to design physical models of phenomena that simulate key aspects of black holes? In what ways are advanced computer simulations of black holes the modern equivalents of "laboratories"?

VIII. Fundamental Questions

The need to stay in contact with experiment

HB: I'd like to ask about using galactic dynamics and our understanding of galactic structure as a probe for some other aspects of fundamental cosmology. We've talked about dark matter a little bit but we haven't talked about things like inflation, dark energy or anything like that.

Might there be ways, with a more detailed understanding of galactic dynamics, to be looking for signals of inflation, or elucidating key issues with respect to dark energy, and so forth?

ST: Well, the main difficulty in doing that is that the physics that we know about—physics of matter or radiation—is really complicated. If you want to understand star formation, you have to understand the relations both between gas on scales of the whole galaxy and on scales of the radius of the star, which is many, many orders of magnitude smaller.

So, paradoxically, the things that we know least about—dark matter and dark energy—are the things that are the easiest to calculate. When we get to understanding star formation, the interaction between stars and the remaining gas, the formation of a black hole in the centre, we really get challenged to understand things correctly.

The problem, then, in using galaxies to probe dark matter and dark energy is that the processes going on in the baryonic part of the galaxy are much harder to understand than the processes going on with the dark matter and dark energy. Where the galaxies are important is simply as test particles, or as probes, or flashlights, to map out the geometry of the universe.

You get tremendous amounts of information simply from the statistics of the large-scale distribution of galaxies. You can use galaxies that are sometimes visible at very large distances to probe how the geometry of the universe changes with distance, and in those cases you're trying to work on scales big enough so that you're not affected by the detailed physics that goes on within the galaxy, but just to use the galaxies as unbiased probes of the structure of the universe on large scales.

The basic assumption, then, is that while the properties of one galaxy here and another over there may be different, statistically, they're both determined by the same processes if the universe is homogeneous on the largest scales.

HB: OK. Let me move on to a more "sociological"-type of question now, related to some previous conversations we've had. The word "fundamental" sometimes gets thrown around a fair amount in the physics community—this question of what is "fundamental" physics and what is not.

In recent years—and by that, I mean a couple of decades—there has been a trend towards people who regard themselves doing fundamental physics being increasingly removed from the realm of experiment, not least of which because often there simply isn't enough relevant data around for their line of work.

I'm hedging a bit, because I'm trying to be respectful, but let me now try to be a little more explicit. You're what I would call a "real physicist," and by that I mean that you not only have a deep understanding of theoretical structure and the corresponding mathematics, but so much of what you've done in the past and continue to do involves what we can *actually* look for, measure, make predictions about and so forth.

Have you been frustrated by some trends in fundamental theoretical physics where some people might spend an entire career without any realistic possibility of actually having any empirical verification for what it is that they're doing?

ST: I think that there is a problem for physics as a whole, a problem which has arisen because, as the saying goes, "experiments consume theory". The theory that was worked on 50 years ago has been tested pretty thoroughly—most of it's right, some of it's not right —but if you want to continue to develop your understanding of physics, you have to develop deeper theories that can only be tested by deeper, and usually more expensive, experiments. I have no problem if my colleagues are doing really good physics that is a long way from experimental tests, but there is, I think, a genuine problem for the community as a whole if that situation continues.

Take the Large Hadron Collider at CERN. If it doesn't discover any new, interesting phenomena in the next few years, it will be hard to know how to design the next, more expensive accelerator; and perhaps hard to persuade the nations of the world to invest billions of dollars in what would be, at some level, a fishing expedition, because you don't have the theoretical guidance to ask precise, well-formed, questions.

That problem is most acute in elementary particle physics, but it's beginning to become acute in astrophysics too. You can still do very interesting, important, fundamental physics with astrophysical missions that, although expensive, are still a much smaller fraction of nations' science budgets.

But there is a real concern that, in particle physics, we won't have new, experimental data that yield really fundamental and puzzling new results that have to be explained, giving us some guideposts as to where to go.

There is an emerging problem in astrophysics that, if we continue not to find the actual dark matter, or if we continue to be unable to understand the nature of the dark energy, then, at some point we're going to have to take a hard look and figure out what to do next because we can't keep building more and more sensitive detectors to look for the same dark matter candidates forever. At some point, we have to say that we've been barking up the wrong tree and we have to do something different.

My own focus on what might be called "non-fundamental physics" is simply because I think *all* physics is interesting, and it's most interesting when it's directly connected with data, with either experiments or observations. I think there are cases where some of the subject areas in astrophysics and particle physics that haven't been connected with data have been much less productive than you might have hoped.

In the 1980s when inflation was developed, there was a real belief that there was this grand new synthesis in physics where the very small—that is, particle physics—and the very large—that is cosmology—would come together and produce really radical and dramatic, new insights. And so far that hasn't happened, in the sense that we haven't really learned anything new about the fundamental structure from cosmology yet.

We've learned some new things—from solar-neutrino experiments, from cosmic ray experiments—but the grand vision that we would detect a particle that's responsible for the dark matter that would provide insight into the fundamental structure of physics and how we should improve the standard model of particle physics, hasn't happened yet.

I hope that it will, but it's true that I have colleagues who have been trying to do that for decades and have had lots of good ideas, but less success than I think they might have hoped for initially.

HB: Since you brought up inflation, perhaps now's a good time to bring up the controversy regarding the BICEP2 experiment and its interpretation, together with the way the findings were initially presented and eventually recanted. Are there any lessons to be learned from any of that, in your view?

ST: I think that the lesson that I took away, or that I was disturbed by, is that, of course, anybody can make a mistake; but there was a lot of talk in the media about saying, *This is the normal process of science in which this is the way science works and this is the scientific process and this is how things are supposed to work.*

I think that's not true: you're not supposed to publish results that are incorrect, and you're not supposed to have press releases about them—so I think this is emphatically the way science *isn't* supposed to work.

The original result was, I believe, a seven standard deviation result, which meant that it was supposed to be correct to the 99.9999% confidence level. And it wasn't.

I think that was a blunder—and anybody who tries to sugarcoat it to make it look like it wasn't a blunder is corrupting the public's view of the scientific process. In the extreme case, people might say, *"Well, they said that there was evidence for inflation and there wasn't; they say there's evidence for climate change, maybe they don't know about that either..."*

HB: People might lose faith in the scientific process, then, or might not have sufficient understanding of what is the *genuine*, scientific process?

ST: I think that's a real concern. Obviously, you can't "put the genie back in the bottle," but at the very least I think the people who made the mistake should admit it and apologize; and I think they haven't been quite as forthright as that.

Questions for Discussion:

1. How would you distinguish between "fundamental" and "non-fundamental" physics?

2. Are too many of today's theoretical physicists "too disconnected" from the experimental realm? If so, is there anything that realistically could, or should, be done about it?

*3. Have you heard of the BICEP2 controversy that Howard and Scott address at the end of this chapter? Do you think the aftermath and analysis received sufficient attention from the media and the general public? Those interested in an additional perspective on this issue are referred to Chapter 5 of the Ideas Roadshow conversation **Inflated Expectations: A Cosmological Tale** with Princeton University physicist Paul Steinhardt.*

4. To what extent do you think that the general public trusts science and scientists?

IX. Concluding Thoughts

Public policy and capitalizing on the moment

HB: Let me pick up on that to ask you a bit about the reputation of science in the general public. Are we, as a society, doing a sufficiently good job generally at explaining what science is, explaining the scientific process and in educating our populous as to how to develop a deeper appreciation of what science is and how integral it might be? And are there places that, in your view, that are doing it better than other places?

ST: Well, that's a highly politicized question, of course.

HB: You don't have to answer. You can just say, "*No comment.*"

ST: Well, we know, for example, that, I think, something like half of the population in this country doesn't think the theory of evolution is correct. In fact, I wouldn't be surprised if it was more than that, actually. I think it's better in European countries, but there's still a significant fraction of people who don't believe in it.

It seems to me that as long as that's the case, that's a pretty obvious, initial sign that we're not doing a good enough job at communicating, not so much the science, but the *nature* of the scientific process: what the difference is between scientific fact and other kinds of facts, developing an educated population that can distinguish science from non-science and make informed judgements about what you should believe and what you shouldn't believe.

HB: So how might we do that better? Might you be able to point towards some concrete measures that might be taken to improve the situation?

ST: There, I think I *will* say, "*No comment.*" This is not a subject that I've thought a lot about. The only thing I'd say is that, of course, lots of people have opinions on how to do things better, but this is actually a question that's susceptible to scientific investigation.

People who do psychology and other sociological disciplines, know how to do this, and I think that the important thing to do would be to treat it as a scientific problem and try to understand, scientifically, what the best way to do this would be.

HB: Let me be devil's advocate for a moment. I've talked to a lot of people who are passionate about the importance of communicating science to the general public, communicating science responsibly, educating people responsibly and so forth. Obviously these are goals that I subscribe to as well, but sometimes I wonder if any of this makes any difference whatsoever.

So this is what I mean, here is my devil's advocate position. For the vast majority of human history, most people have been incredibly ignorant about all sorts of things, certainly including science. Nonetheless, while there have been periods of highs and lows, of course, over the last 300 or 400 years, there has been a steady, consistently monotonically increasing level of understanding and awareness which has developed across society regarding basic science.

And that has happened in spite of the fact that there has *not* been, at any time that I'm aware of, an overwhelming endorsement of the scientific process by society at large, or huge, overwhelming amounts of money being devoted to scientific endeavours, other than the odd specific case here or there.

What I'm saying is that, by and large, science or scientific awareness do not feature largely in public elections or media headlines or, more generally, in the public consciousness; and so one might think that none of that really matters much at the end of the day. Science will keep doing what it's doing through the efforts of a small, dedicated minority such as yourself, who are doing what you're doing simply because you're passionately driven to understand the world around you.

There are people just like you who are coming through the ranks now, and it will be ever thus: you will never be a majority, but you will be actually driving society forwards in some non-trivial way and that's just the way it is and we shouldn't wring our hands or worry too much about it.

ST: Well, I think the counterargument to that is that science now plays a much larger role in life than it did 100 or 300 years ago. A much larger fraction of the questions that politicians and society have to deal with are scientific ones: global warming, pollution and the environment, biomedical issues—how to deal with Ebola, how to improve the health of the population as a whole—to even economic issues, such as how to keep the financial system stable.

Far more of these questions require scientific answers—or at least informed, technological answers—than was the case 100 or so years ago. So, I think it is more important than it used to be simply because science has become a more powerful and bigger part of our lives.

The second part of your devil's advocate view seems to me to be really more a question about support for science and science funding. Now, I believe that science has been one of the main drivers of economic growth in this country and most others. There are studies that agree with that—which I'm not an expert on, so I won't try to go into them—but my personal belief is that science should be funded by a larger fraction of the national income than it currently is.

Having said that, I have some sympathy for a different point of view than many of my colleagues who say, "*Science is not adequately funded and we're going to push for a bigger science budget in the next fiscal year.*"

While I agree with their overall motivations and an increase in science funding generally, it doesn't address one of the major, structural problems in the scientific enterprise, which is that much of the enterprise is built on the assumption that you're going to have exponential growth: the relative numbers of graduate students, postdocs, people supported on soft money, faculty, national labs—all of these things are built on an integrated model of the life-cycle of a scientist or a scientific project or a scientific field that are based on the assumption of exponential growth.

And I think what the community has to ask itself is, *"What's a viable or optimal model of the scientific enterprise and the scientific community in a world where there is zero, real growth?"*

Until you have that vision in mind, and until you have a viable model in that form, putting your efforts into getting a 5% increase instead of a 3% increase in the next year's budget is not going to be very helpful, because if you *do* succeed, as long as you're assuming exponential growth, all that's going to do is delay the reckoning by another year or two.

So, I think the community needs to think harder about how many graduate students they should train. Is it okay for graduate students to go work in the data industry or the financial industry? And, if so, is the best time to do so after a bachelor's degree, a Master's degree, a PhD or a postdoc? What should be the ratio of postdocs to permanent positions? What's the optimum distribution between funding large projects and small projects? What's the optimum distribution between targeted grants and curiosity-driven grants? And all of that should be done in a context or a framework of zero real growth; and that has not yet been happening.

HB: Is there some wider sympathy for this view, or are you a lone wolf out there in believing that this should be done?

ST: I think that there are some straws in the wind that say that this point of view is taking hold. There was a very influential paper recently in the biology community arguing that our current model for funding research is broken and recommending some changes along these lines, but I think it's not a general point of view yet. I think it's a point of view that people have been tiptoeing around but not addressing head on.

HB: To finish up, I'd like to return to science and ask you what's keeping you up at night. A standard question that I tend to ask is: *If I were some omniscient being and I could provide you with the answer to any question you should ask me, what would you ask?*

ST: What is the nature of dark energy? What is the nature of dark matter? How can we unify quantum mechanics with general relativity? That was easy.

HB: Yes. I was hoping for something else.

ST: You were hoping for something easier?

HB: Yes.

ST: Well, you only gave me three. If somebody's omniscient, you shouldn't waste your opportunity.

HB: Fair enough. OK, I'll give you two more.

ST: What was the chemical mechanism for the origin of life on earth? How common is life on other planets, and what form does it take?

HB: Goodness, you really take no time at all—most people at least pretend to ponder these things a bit. I should give you another five, but we should probably stop. Is there anything we missed? Anything that you'd like to add that we haven't had a chance to get to?

ST: No, I think you've pretty much covered everything.

HB: Well, it's been a lot of fun, Scott. Thank you.

ST: My pleasure.

Questions for Discussion:

1. Do you agree with Scott that the scientific community should restructure their research environment to accommodate a world of zero real growth?

2. Is there a pressing need for an increased number of "scientifically literate" politicians to deal with the increasing number of scientifically-related issues that need to be addressed politically? What, in this case, does "scientifically-literate" actually mean?

3. How can we increase the interest and understanding of science throughout our society?

4. What do you think Scott means, exactly, when he talks about the ability to distinguish between "scientific fact and other types of facts"?

Continuing the Conversation

Additional Ideas Roadshow conversations not offered in this collection that the reader might enjoy include *SETI: Astronomy as a Contact Sport* with astronomer and former SETI director **Jill Tarter**, *Pushing the Boundaries* with Institute for Advanced Study scientific polymath **Freeman Dyson** and *A Matter of Energy: Biology From First Principles* with UCL biochemist **Nick Lane**.

A Universe of Particles:
Cosmological Reflections

A conversation with Rocky Kolb

Introduction

The Passion Principle

Say the word "scientist" and the picture that comes to mind for most people is a bespectacled, dispassionate type, plodding along in a carefully rigorous, decidedly unexciting, way to test some specific hypothesis or conjecture. Those of a certain age will conjure up images of an unimposing fellow with a slide-rule jutting from his belt, while others will envision a pocket calculator proudly on display, but while the technology changes, the essential stereotype does not: scientists are essentially nerdy, Spock-like types who methodically go about their business in a quiet, matter-of-fact sort of way, deliberately eschewing any great outburst of sentiment or emotion.

Well, meet Rocky Kolb.

Upon first encountering this lanky, gregarious Louisianan, you likely wouldn't be surprised to learn that he was anything from a stand-up comedian to a chief sales executive to a former professional basketball player. But Rocky is none of those things. He is a highly accomplished theoretical physicist who specializes in applying abstract mathematical concepts of modern particle physics to the early universe.

And he is also someone who has made an entire career of being strongly guided by his emotions, right from the start.

> *"I grew up in New Orleans, Louisiana. In the dog days of summer, we'd play baseball in the morning, but by noon or so it was simply unbearable to be outside—or inside for that matter. The only place I could walk to that was air-conditioned was a small public*

*library branch. So to be cool, I went to the library. And when I got to the science section of this little library, I remember the librarian coming over and pulling me away from the science section, saying, '**No, you want to go to the children's books section**'.*

*"I had already read those books—they didn't particularly interest me—but the science books, they were **forbidden**. And every time she wasn't looking, I would run over and get a science book, and then put one of these big golden books outside of it, so I'd be reading science when she wasn't looking.*

"And it turned out that I really loved it; and the part that I really loved, were the parts about physics and astronomy. I was fascinated with both the smallest things that people knew about, and also the largest things—in astronomy. And that sort of stuck with me."

Of course most people have a nice, heartwarming story to tell about how they began their careers—a good tale to tell your grandchildren about, if nothing else. But for Rocky, his strong, emotional attachment to both the intellectual appeal of physics and the warm, collegial atmosphere of the scholarly community only continued to strengthen.

*"In the late 1970s I became interested in neutrinos. Could neutrinos have mass? Could there be any number of neutrinos? And my advisor, Duane Dicus, said, '**Well, doesn't cosmology or astronomy say something about neutrinos?**' This was a really great thing. He didn't know any cosmology, I didn't know any cosmology, I'd never taken a course in cosmology. I'd never taken a course in astronomy or astrophysics.*

"And so we learned it together—I tell people that, I picked up astronomy on the streets. He was a little bit faster than I was, but it was a great experience to learn something with my advisor. I think this was a much better experience for me than having an experience where the mentor already knows something about the subject and you just try to absorb it. It was really a great learning experience."

And then, a few years later, as a postdoc at Caltech working with the Nobel Laureate Will Fowler, there was an even stronger emotional experience:

> "*Perhaps he greatest thing I learned from Willy was to appreciate the work done by your students and postdocs. It's rarer than you might think: to really appreciate and celebrate what your students and people who work for you are doing. He always showed real enthusiasm. He wasn't involved in the papers: he didn't really know the science of particle physics very much, but he really loved it; and he instilled in me and many of the people who worked there, really, an idea of a sort of 'family pride': we were 'Willy's boys'.*

> "*I remember being at a conference somewhere years later and a bunch of Willy's former postdocs and students were sitting around talking, and there was a senior professor from another institution—I won't mention his name—made some comment to the effect,* **'Well, of course, Willy did all these things because his postdocs and students were so good'***, or something like that. It was sort of like,* **'What did he do?'** *You know,* **'He was just at Caltech. He just got great students and postdocs'.***

> "*And I truly feared for his life. There were several of us who got very red in the face and looked dangerously close to strangling him.*"

Well, OK, you might think to yourself, I'll grant you that in any professional community passions can run high when it comes to matters of internal sociology and protecting vested interests. One can even imagine—albeit dimly—that different schools of chartered accountants can be roused to near-passionate expressions of tribal loyalties. But the actual ***practice*** of something like theoretical physics—surely that's a domain where any strong emotions are left at the door.

But yet again: no. Here's Rocky talking about dark matter, for example:

"Personally I love this idea of an elementary particle to explain dark matter, which holds galaxies and other large scale structures together, because my original loves when I was a 10-year-old boy were astronomy, the big things, and particle physics, the little things.

I had never thought—nobody, I suppose, had thought—that there might be a deep and profound connection between the two. So it's wonderful that I just came along at the right time.

And here he is talking about dark energy:

*"Dark energy, to me, is like fingernails on the chalkboard. It just drives me nuts. I don't like it. I don't like it; I admit that it's a prejudice, but there it is. I don't have a good explanation. It's not a logical thing. It's not that I say the observations are wrong— although I did for at least a couple of years. I kept saying, '**There must be some other effect responsible for these observations**', until I was finally convinced otherwise. I just can't swallow it. I think we all have these prejudices, these ideas, that we just can't swallow."*

Like Rocky, most scientists don't like to admit that their professional judgements are influenced by their emotions. At some level they feel they should be able to somehow transcend such base sentiments and be able to look calmly and objectively at the facts that are presented to them. But that is not just impossible—because science is a human activity done by humans—it is often very far from desirable. Many of the most impactful scientific accomplishments, for example, were made by those who were motivated—often very strongly motivated—by a deep and abiding aesthetics sensitivity: a quest for beauty, for harmony, for simplicity.

And then there's the simple and obvious fact that scientific success, like success in any highly-challenging domain, requires a level of dedication and commitment that naturally springs from the deepest wells of personal desire and passion.

It was all of a piece, then, that when I asked Rocky what advice he would give to science teachers, he had this to say.

> "If I could sum it up in one word it would be **'passion'**: have a true passion for what you do and convey that to the students, trying to ignite a passion in them. Don't be so preoccupied with the specifics of the curriculum and all these other things: it's the passion that matters most. Nothing else matters."

The Conversation

I. Cool Beginnings

From the local library to Caltech

HB: Perhaps you could talk a little bit about your youth: how you got interested in science, how you got interested in physics in particular, and whether or not there were other scientific interests that you had growing up.

RK: Well, I grew up in New Orleans, Louisiana. And back then—in New Orleans in the 1960s—air conditioning was very rare: our house wasn't air-conditioned, and it was hard to find a place that was.

In the dog days of summer, we'd play baseball in the morning, but by noon or so it was simply unbearable to be outside—or inside for that matter. The only place I could walk to that was air-conditioned was a small public library branch. So to be cool, I went to the library.

It was a very small library, so I decided to systematically go about trying to read every book in it—I didn't have anything else to do. And when I got to the science section of this little library, I remember the librarian coming over and pulling me away from the science section, saying, *"No, you want to go to the children's books section."*

HB: How old were you at this point?

RK: About 10 or so, I think. I'm not exactly sure. At any rate, I had already read those books—they didn't particularly interest me— but the science books, they were *forbidden*.

HB: So she piqued your curiosity.

RK: That's right. And every time she wasn't looking, I would run over and get a science book, and then put one of these big golden books outside of it, so I'd be reading science when she wasn't looking.

And it turned out that I really loved it; and the part that I really loved, were the parts about physics. For some reason, neither the biology or chemistry parts really interested me, but I was strongly attracted to physics and astronomy. Astronomy is sort of a common interest among small children, and physics maybe less so, but I was interested in things like the nucleus and the inner workings of things: *What are things made of?*

I was fascinated with both the smallest things that people knew about, and also the largest things—in astronomy. And that sort of stuck with me.

HB: So your interest was piqued then. Was it fostered at school, or did you have to fight through some level of indifference in those surroundings?

RK: Well, in elementary school, I typically felt that I knew more than the teachers in science and math, and it really wasn't interesting to me at all—but that was true of many subjects in elementary school. I just didn't pay attention and didn't have any problems with the classes. High school was a little more challenging, but again, in science, I didn't have any problem. And mathematics didn't particularly interest me at the time. I was sort of a B-student in math without any effort, but later when I started studying physics more seriously and began to use mathematics, then I became very interested in it.

HB: So this fairly standard view that mathematics is a tool for you to understand physics.

RK: Yes, it was a tool.

I should back up a little bit and say my interest in mathematics may not have been the mathematics that was taught in school, but my father was interested in cryptography and mathematics.

So, although I was interested in physics and astronomy early on, mathematics did not really attract my attention; and I became interested and really talented in mathematics through taking physics, once I could see what the mathematics was used for. "*Ah,* **this** *is how you use algebra.* **This** *is how you use trigonometry or calculus.*" It was a tool, and I was not particularly interested in the mathematics that was taught at school.

My father did not have much of a formal education, but he was interested in mathematics, and he had a lot of books on code-breaking, cryptography, and sort of abstract ideas in mathematics that I read and we talked about. So I was interested in the abstract ideas of mathematics, but the everyday grind of the multiplication tables, or as we called them in New Orleans, the "gazintas": Eight gazinta sixteen twice. It was a New Orleans things, the *gazintas*.

Or even algebra. To me, it was sort of drill and toil, rather than anything intellectually exciting. I could catch the intellectual excitement of physics and astronomy, but I couldn't catch the intellectual excitement of mathematics until I started using it more directly.

HB: Other than cryptography, did your father have any other mathematical interests? Where did his curiosity about cryptography come from? Do you know?

RK: My father was an older parent. When he was 12 years old, his father died in the 1918 flu epidemic, the Spanish flu. He was the oldest child, and his mother told him, "*Okay, now you have to support the family*".

He had four siblings. I can't imagine being 12 years old and being told you have to support your mother and the rest of the family, so he had to drop out of school. He later went and completed it at night and continued to educate himself at night school, but he was put to work from a very early age.

HB: That's a tragic story.

RK: Well, in some ways it is a tragic story, but you know, he was never resentful about it. Never resentful.

HB: And presumably he somehow managed to develop interests in many other things, such as cryptography.

RK: Although he didn't have much of a formal education, the house was full of books. We didn't have a television or a radio when I was a child. He would read Shakespeare; so even before I started school I didn't have anything to do, so I started reading as well.

HB: Well, it seemed to have worked out well. Maybe that should be invoked as social policy. Shut off the media and the air conditioning, and force people to go to the library.

RK: Right. And it should be *forbidden* to study science.

HB: You talked about elementary school, and there don't seem to have been any teachers there who had a significantly positive influence. Did that change at all when you went to high school?

RK: Yes, I had a great high school physics teacher; after that I knew I really wanted to do physics—astronomy wasn't taught in the school I went to.

HB: So your initial motivations combined with the influence of a great high school physics teacher made you say to yourself, "That's what I'll study in college".

RK: Yes. By that time science and mathematics were the easiest courses in high school for me, so I said to myself, "*Well, I'll do something that's easy, that I like*".

HB: OK. And when did you start becoming more and more interested in cosmology and astrophysics?

RK: Well, it's sort of a funny thing. I was always interested in what is now called particle physics and astronomy. But until recently, they were completely separate. As an undergraduate, I took all the physics courses I could.

But I didn't take an astronomy course in college, because astronomy, then, was more descriptive, and I wasn't really interested in the descriptive part of it: finding constellations and things like that. It's fine and all, but to me it's not intellectually exciting to find constellations.

HB: It's like botany, in a sense.

RK: That's right. So I decided, *"Well, I can't do astronomy, I can't do anything to do with astronomy, I will be a particle physicist."*

I went to graduate school at the University of Texas and worked for particle physicists at the Center for Particle Theory there. Early on realized that I wasn't cut out for experimental work. I enjoyed the mathematics and I found that easier than the experimental work—which I respected, I appreciated, but I found sort of tedious.

Of course, it's not if you do it for a living, but the way the labs were in college and high school for me were sort of tedious; and I was impatient to understand *What does it mean?* I was much more interested in the concepts and the ideas than actually using my hands.

So for me, the experimental labs—and this may be an indication of where I went and maybe they weren't very good— were sort of like a cookbook: *"You do this, you do that, measure this,"* and things like that. It didn't seem nearly as attractive as the intellectual challenge of sitting back and trying to figure out how things work.

Anyway, so there I was in graduate school in particle theory at the University of Texas. This was in the late '70s—'77 and '78— and just by happenstance I became interested in neutrinos. Could neutrinos have mass? Could there be any number of neutrinos?

And my advisor, Duane Dicus, said, *"Well, doesn't cosmology or astronomy say something about neutrinos?"*

This was a really great thing. He didn't know any cosmology, I didn't know any cosmology, I'd never taken a course in cosmology. I'd never taken a course in astronomy or astrophysics.

And so we learned it together—I tell people that, *I picked up astronomy on the streets.* He was a little bit faster than I was, but it was a great experience to learn something with my advisor. I think this was a much better experience for me than having an experience where the mentor already knows something about the subject and you just try to absorb it. It was really a great learning experience.

HB: Presumably you could really see, first-hand, his level of honesty and openness: not knowing certain things and not pretending to know them.

RK: That's right: he was very honest about the situation: *We're learning this together, let's figure this out.* Of course we both made mistakes on the way, and he was almost always ahead of me. Occasionally I would pass him a little bit in one area, and I'd feel really good, but then he would move past me again.

Eventually we managed to write some interesting papers, but at this time cosmology was not considered respectable. Astronomy was something different, but particle physicists in particular had—well, disdain's too strong a word—

HB: Really? I think disdain fits perfectly.

RK: Well, their view of astronomy was something like, *"Oh, it's just about orders of magnitude. Nothing's rigorous."*

HB: Hand-waving.

RK: Exactly, it was thought of as hand-waving. And cosmology was even worse: *"How can you say anything about the universe a second after the bang? Where's the experimental evidence?"*

And in some sense they had a point, but I think it was one of those things when there were really no developments because it was thought not to be respectable, so no one really worked on it, so there were no developments—it became something of a vicious circle. I think that there were opportunities to do some things that were just not taken advantage of because relatively few people thought it was a good thing to be working on at the time.

So had I been more intelligent or aware, I would've said to myself, *"This is a terrible career move: to be interested in a field that no one cares about".*

HB: But it turns out that you were brilliantly prescient.

RK: Right, right. I'd like to say, *"No, I had it all figured out"*, but it just worked out that way.

HB: I briefly touched on this during my conversation with Paul Steinhardt (*Inflated Expectations: A Cosmological Tale*)—it was something I hadn't really thought about before, but the cosmic microwave background (CMB) was discovered in the '60s, right?

RK: Yes, 1964.

HB: And it took a really long time before people started to think about using it as a tool, before people thought, "Maybe we can actually look at this stuff a little bit more closely, look for inhomogeneities, look for *something...*"

RK: You could probably count on maybe one or at most two hands the number of people who were really interested in the CMB.

HB: So why do you think that is, exactly? I remember when I was a student there was very much this sense that cosmology was not much different from metaphysics, that it was something to stay away from.

I remember there were various different models of the universe—Bianchi-this and Bianchi-that—but there was this general view that, "*Well, we'll never be able to decide which one we're in, and so all we can do is just enumerate the mathematical possibilities*",—kind of like putting angels on the heads of pins.

That was my sense of the way it was, but now when I look back I can't help wondering why people didn't stop to take a closer look and say, *Hey, there is this CMB that's out there. Why don't we actually look at it more carefully? This could be really, really interesting.*

I mean, I appreciate that there were enormous technical challenges involved in being able to measure these tiny inhomogeneities, but it's like nobody was even thinking that way at the time.

RK: Yes.

HB: So why do you think that was?

RK: Well, I think it's reflective of a sort of prejudice in the field that there was nothing to do there and so it wasn't considered a respectable thing to be doing.

This was true well before the CMB was discovered—in fact, it could have been discovered years before that: there were predictions and ideas that people either didn't know about or didn't take seriously enough: it could've been discovered maybe 5–10 years earlier, depending on the technology.

HB: And maybe even deliberately, instead of by accident.

RK: Yes. That's right, that's right.

At any rate, for me I got my PhD in 1978, and then the usual thing is to take a postdoctoral position someplace; and I didn't think I would go into cosmology or astronomy or astrophysics.

HB: Even then?

RK: Even then. And I received a couple of postdoc offers. All of them except one were in particle physics groups—weak interaction phenomenology, neutrino physics, things like that—but not astronomy and astrophysics.

HB: Of course, this was also a very exciting time in particle physics.

RK: Yes, yes. The Standard Model was just emerging at that time. But there was one offer for a postdoc at Caltech working with Willy Fowler, a Nobel Laureate who received his Nobel Prize for work on the development and application of nuclear astrophysics.

The application of nuclear physics to astrophysics started with Hans Bethe's paper about how the sun shines in the '30s, then it picked up in the '40s and '50s, and the 1960s were the heyday of nuclear astrophysics with Willy Fowler at the forefront of it all.

If you look at the postdocs and students Willy had at Caltech, it's really just amazing—just about everybody went through there. Perhaps the greatest thing I learned from Willy—I didn't actually learn much physics from him honestly—was to appreciate the work done by your students and postdocs. It's rarer than you might think: to really appreciate and celebrate what your students and people who work for you are doing.

HB: How would he do that, exactly? What sorts of things would he do?

RK: Well, it was just a pride. Willy referred to us as "his boys". Now, I should also say that it wasn't about gender—he hired the first women at Caltech—but we were "Willy's boys".

HB: Right: the emphasis was on the "his" rather than the "boys".

RK: Right. It was always "Willy's boys". Whenever we did something and come talk to him about it, he would get the biggest kick out of it and say, "*Oh, that's wonderful. Tell me more about it*".

He always showed real enthusiasm. He wasn't involved in the papers: he didn't really know the science of particle physics very much, but he really loved it and he instilled in me and many of the people who worked there, really, an idea of a sort of "family pride". We were "Willy's boys".

HB: Is there still a sense of that even now?

RK: Yes, there's a group of us: Dick Bond, Stan Woosley, Craig Wheeler. John Bahcall was also one of Willy's boys. Craig Wheeler. It's just a long, long list of people.

I remember being at a conference somewhere years later and a bunch of Willy's former postdocs and students were sitting around talking, and there was a senior professor from another institution—I won't mention his name—made some comment to the effect, "*Well, of course, Willy did all these things because his postdocs and students were so good*", or something like that.

It was sort of like, "*What did he do?*" You know, "*He was just at Caltech. He just got great students and postdocs*".

And I truly feared for his life. There were several of us who got very red in the face and looked dangerously close to strangling him.

HB: So you look to him as an inspiration for group-building.

RK: Yes, that was definitely an inspiration for me. I was the head of a group at Fermilab in theoretical astrophysics for many years and there were many people who came through during that time. I worked with many of them, and hopefully some of them learned some physics from me, but on the whole I probably learned more from them than they did from me.

But one of the things I'm most proud of is the impact I hopefully had on them: how I might have helped their careers and celebrated what they accomplished. I think that's something that's rarer in science than you might imagine.

HB: Indeed. Have you ever gone to a conference when you overheard someone saying, "*Oh, Rocky Kolb, what has he ever done? He's just got good students and postdocs*".

RK: Well, I would hope my students and postdocs would stand up for me.

HB: So what were you working on when you were one of Willy's boys at Caltech?

RK: Well, the postdoc I had at Caltech was one of these positions where you can do whatever you want, so I went there and said to myself, *Well, I'm sort of interested in cosmology and astrophysics. I'm also interested in particle physics. I'll take a few months, go to a couple of talks, talk to people, and sort it out.*

Because I thought, *Well, now I'm a postdoc, I'm no longer a graduate student. I have to get serious. I have to choose a direction, something to do in my career.*

During the first month or two I was at Caltech, there was a series of lectures given by a famous cosmologist, Allan Sandage. He was Edwin Hubble's student and assistant and to some extent carried the mantle of Hubble, the father of modern cosmology.

So I went to these lectures. This was old-time cosmology: all this nomenclature and so many graphs that looked like random dots that he drew straight lines through without arrow bars. I thought is was just terrible.

So I said, "*Forget it. Nobody knows anything about cosmology. I give up. I'm going to be a particle physicist*". Then a couple of weeks later, there was a lecture by someone at Caltech. I won't mention his name: I'll tell you off camera. It was on string theory. This was 1978 or '79.

HB: I think I have an idea.

RK: This was in the very early days of string theory; and after the lecture, Murray Gell-Mann got up—you know, the great Murray

Gell-Mann—and said something to the effect of, "*This is the future of particle physics*". So I thought to myself, *Well, maybe cosmology wasn't that bad.*

HB: There's always room at the bottom.

RK: Well, this was not modern string theory.

HB: This was strong interaction stuff, right?

RK: Well, it was sort of strong interaction stuff, but I think Murray Gell-Mann saw that it had the potential to be a fundamental theory and really apply to gravity. Other people too were just starting to realize that. But it was so technical, and seemed so removed from anything, so I thought to myself, *Well, maybe cosmology wasn't so bad. If that's what I have to do to be a particle physicist, I'll be a cosmologist.*

But luckily I was in an environment as a postdoc where I could do that; and also the career path in those days, I think, was easier or less structured than the career path nowadays.

I ended up getting positions and doing okay, but it was during a time when there seemed to be more latitude to try different things. Nowadays I see graduate students and postdocs who are so focused—"*This is what I want to do, this is my career path*"—and they are so determined to follow that narrow trajectory.

HB: So as you're talking, it's clear that there's a significant, impactful sociological aspect to this in terms of how science develops when the people coming into a field are so narrowly focused and are fixated on following the leader and doing whatever is considered the hottest or trendiest sort of thing to be working on at the moment and all of that.

But it's perhaps also worth remarking on a more personal practical factor associated with a sense of general broad-mindedness, and concrete experience in an array of different areas.

Here you are talking about your graduate student and postdoc inclinations, thinking about moving from particle physics to cosmology, at a time when cosmology was far less popular than it is today.

As you say, that attitude seems much more prevalent in your generation. Again, as you know, I just spoke with Paul Steinhardt, and he is clearly interested in so many things, from particle physics to cosmology to material science.

And I can't help but think that people who have experience in a variety of different research areas and problems are not only more broad-minded in a sense, they also have a greater ability to solve problems because they have a more robust toolkit of techniques and concepts and are able to make analogies and parallels with other situations.

RK: Yes: that's really helped me. As I said, my background was in particle physics and I was fortunate enough that this field really started around me—with me hopefully playing a small role. But if I had written an application for a postdoc or for a job saying, *"I want to do particle cosmology"*, people would not have known how to read it and it would have ended up immediately in the wastebasket. In "the old days", up to the '70s and '80s, it used to be that someone of the stature of Willy Fowler could write a letter and say something like, *"This guy's really smart, you should hire him; I don't know what he's going to do but you should hire him because he's really smart"*.

And that was one of the advantages of "the old boy network". Now, it also had many disadvantages for which it should be justly criticized, but that was one of the advantages.

Too often today, when we hire people, we do very narrow searches—*We're looking for someone who works in 12.3 dimensions on this problem and applying this technique*—instead of taking a much broader view.

I always try to encourage people to do broad searches, and occasionally take a chance: roll the dice, you know? Keep an eye out for people who look like they're changing fields, or starting in a field that's not very popular, because when you work in a new field, there often aren't many senior people around to write letters and fully appreciate what you're doing.

HB: Which not only makes your own research group more dynamic, but is very important to the future of the entire field as well. Let me try to briefly sketch out what I mean.

You mentioned Willy Fowler, a hugely accomplished nuclear astrophysics guy in the 1960s. And then with the Standard Model of particle physics getting established in the 1970s, it's hardly such a big stretch to go from nuclear astrophysics to particle astrophysics as our understanding of particle physics rapidly develops.

Of course predicting how, exactly, this will occur is another question entirely, but the basic structural relevance seems quite obvious—but only if you deliberately make an effort to get a big-picture view of things.

RK: I think Willy saw it, I think he saw it. Willy was very excited about it, and he used me to teach him the Standard Model, to teach him the Weinberg-Salam model. He wasn't about to go to Dick Feynman and Murray Gell-Mann and say, "*Explain this to me*." But he could come to me.

It's just like Duane Dicus, my PhD advisor at Texas: he wasn't afraid to say, "*I don't understand this. Explain this to me*".

Questions for Discussion:

1. Do you think that theoretical physics has a greater tendency to be fashion-driven than other fields of science?

*2. Are you surprised at the extent that many young physicists are preoccupied about their next job rather than simply pursue subjects that they find intrinsically interesting? What effect do you think this has, generally speaking, on not only our collective understanding of physics but also on the type of person who becomes a professional physicist? (Readers interested in this topic are referred to Chapter 10 of **Pushing the Boundaries** with Freeman Dyson when he highlights the economic factors associated with modern string theory.)*

3. In what ways has the popular legacy of Einstein influenced the way physicists think of themselves?

II. Cosmic Inflation

Alan Guth causes a stir

HB: So let's talk a bit more about your experiences as being front and centre during this sudden explosion of interest in modern cosmology. Tell me what that was like.

RK: I was lucky enough to be part of the group of people who developed the so-called Standard Model of cosmology. We didn't know what we were doing. We didn't have any grand designs in mind, and it was "a children's crusade": there were really no senior people in the field.

Now, I'll mention one person whose work I think went a long way to legitimizing it: Steven Weinberg. Steve wrote some of the early papers about cosmology, as well as wrote a famous book, *Gravitation and Cosmology.*

HB: But that was mostly gravitation—

RK: It was mostly gravitation, but he was clearly interested in cosmology even though he didn't really work very much in that area. He wrote papers on neutrinos as dark matter, which were personally very influential to me; and he also wrote a popular book, *The First Three Minutes*. So I thought to myself, *If Steve Weinberg does it, it can't be too terrible, right?*

HB: He legitimized it, to some extent, by his presence.

RK: That's right He was just about the only senior, established person to work in that field at that time. I mean, he started in the 1970s.

HB: But other than him, as you were saying, you and your colleagues were part of this "children's crusade".

RK: It was a wave. I was caught in the wave. It was a very exciting time, because there were all these new ideas and possibilities: phase transitions, inflation, dark matter, cosmic strings. It seemed like you could wake up in the morning and have a new idea and write a paper that afternoon.

HB: Whatever happened with cosmic strings, anyway?

RK: Well, they've been abandoned as leading to large-scale structure: seeding galaxies and so forth.

HB: Why?

RK: Because they don't predict the fluctuations seen in the CMB.

HB: That's a good reason.

RK: This is a good example of how experiment and theory come together. The Cosmic Background Explorer (COBE) was a satellite in operation from 1989 to 1993 built to carefully study the CMB. It discovered these tiny fluctuations in the microwave background radiation and after COBE they started mapping out the details of the fluctuations with greater and greater precision.

Cosmic strings made one general prediction and perturbations from inflation, or seed perturbations, made another prediction; and they were strikingly different, and the experiment settled it. So cosmic strings did not seed the growth of galaxies, and clusters of galaxies, and other large-scale structure.

That was the reason they were originally proposed—or much of the motivation—but the idea is still around and we could have a "cosmic string renaissance". Maybe superstrings are like cosmic strings somehow. Or maybe they're around, but don't do that, but do something different.

Look at string theory today. Originally it was proposed as a theory of the strong interactions, and that didn't work. But then it was realized, *"Well, it doesn't work there, but perhaps it can work as a theory that incorporates gravity"*.

HB: OK, but let's get back to your story, your "children's crusade". In particular I'd like to get your story about inflationary cosmology: how it developed, your judgement of its current strengths and weaknesses and your assessment for its future prospects.

RK: Well, I think the idea was in the air at the time. Many of us are kicking ourselves that we did not come up with the idea of inflation. There are some arguments about who, exactly, was the first person to develop the idea, but to my mind, the first time I really saw it crystallize in a theory was in Alan Guth's very influential paper.

As it happens, I had been working on similar things, thinking about similar things, while at Caltech with a graduate student there, Stephen Wolfram.

HB: Oh really? The *Mathematica* guy.

RK: He went on to do *Mathematica*, that's right. Stephen was a very young—

HB: Wasn't he like 12 or something when he got his PhD?

RK: Well, I think he was about 16. He wasn't quite shaving yet. I remember him driving—he didn't do it very well; it used to scare the hell out of me—but I do remember him driving. So he must have been over 16 but he was definitely under 18. Anyway, we were talking about it: we had this idea of a phase transition leading to an exponential, a very rapid, growth of the scale factor of the universe. We didn't really understand what we were doing—like Alan did—but we made a terrible mistake. We went and talked to Richard Feynman about it.

HB: Oh really?

RK: It was a terrible mistake. He immediately pointed out the problem with the original idea of inflation: *How do you end it?* That was the problem in Guth's original paper, the problem that the work of Steinhardt, Albrecht, Linde, and others solved.

And we said, "*We don't know.*"

So he replied, "*Well, then, it doesn't work. It might be an interesting idea, but it doesn't work.*"

In hindsight, if we had been as smart as Alan Guth, we would have recognized the genius of the idea and, while admitting it doesn't work, deciding to put it out there anyway.

HB: Well, that seems like more of a tactical thing, it seems to me. I mean, it depends on how you define "smart" I suppose—

RK: No, no, I would say it's smart. It's genius. Alan had a much deeper appreciation of what it could do than we did. Now I always warn people, "*Don't listen too much to senior people.*" Whenever they come to me and ask, "*Do you think this is a good idea?*" I'll say, "*Maybe yes, or maybe no. But don't listen to me.*"

HB: Because even if your criticism is right, it doesn't necessarily mean that's the end of the story.

RK: Right, right. That's right.

HB: And Feynman saw that immediately?

RK: Yes. I'm told he was a clever guy.

HB: OK, so back to the story. Alan Guth realizes the problem Feynman immediately points out but publishes his paper anyway. So then what happens?

RK: Well, inflation is a great phenomenological success without an underlying theory or understanding of what's at the guts of it. Let me be just a little bit more technical there.

HB: Sure.

RK: There is a quantum field, a scalar field, that people say drives inflation. It's called the inflaton.

HB: Not coincidentally.

RK: It's a great name. You have to have a great name. But who *is* the inflaton? *Where* does it come from? I work a lot on inflation, but I won't be completely sold on the idea of inflation until there is some fundamental theory that the inflaton is naturally embedded in.

HB: Right. Because it seems rather ad hoc otherwise, right?

RK: Yes. But on the other hand, until a couple of years ago I would say, "*Well, everything depends upon this scalar field, but by the way, we have never discovered a fundamental scalar field*". But then we discovered the Higgs, so perhaps there *are* fundamental scalar fields.

But for 30 years people were writing models of inflation based upon a type of particle, a scalar field, of which there was no example in nature. And that gave cause for concern. Then the Higgs was discovered, so maybe it's not such a terrible idea after all.

I think that there's something about the theory of inflation that will survive. It may not be in the same form that we have it now, but there's something about it that I think will survive.

Questions for Discussion:

1. *Do you think that Rocky is right to be sceptical of the theory of inflation unless there is "some fundamental theory that the inflation field is naturally embedded in"? To what extent might two physicists differ on what it means for something to be either a "fundamental theory" or "naturally embedded" in a theory?*

2. *When should scientists publish an idea that they know to be wrong at some level?*

3. *To what extent do you think the theoretical physics community would share Rocky's view that the theory of cosmic inflation has been "a great phenomenological success"? (Readers interested in contrary views are referred to Chapter 4–5 of **Inflated Expectations: A Cosmological Tale** with Paul Steinhardt and Chapter 3 of **The Cyclic Universe** with Roger Penrose.)*

III. Dark Matter

Finally recognized ignorance

HB: We talked a little bit about inflation. There are these other two big elephants in the room: dark matter and dark energy. As somebody who's been at the forefront of this field during this extremely exciting time of so many rapid developments, I'd like you to talk about the field more generally and then move on to give me your views on both dark matter and dark energy.

RK: Nowadays we often talk about the "standard cosmological model", and paper after paper invokes this idea. In some sense, it's remarkable: you can account for a great many observations on the basis of this standard cosmological model; and I never would have imagined 20 or 30 years ago being in a position where we could have a model, coupled with the observations that we do, that accounts for large scale structure, distribution of galaxies, microwave background and so forth.

There are a *tremendous* number of observations that either can be explained by the Standard Model, or—perhaps the calculations are too hard to do right now—but are certainly compatible with the Standard Model.

HB: And to go back to what you were saying earlier, when you talked about those presentations by the likes of Allan Sandage with all those dubious graphs and the sense so many people had of the field being equivalent to a sort of "hand-waving metaphysics", to have moved in such a short time to having become one of the most precise, data-driven, fields of science in such strong correspondence with observation—well, it's a truly remarkable transition.

RK: It has been a breathtakingly rapid transition from marginal to respectable.

HB: More than just "respectable", in fact.

RK: That's right: even more respectable, sort of as a leading idea.

Going back to my own experiences, if somebody asked me what I was doing in 1979, I would admit to being a cosmologist, but would beg them not to tell anyone—it was like, *What if my mother finds out that I'm doing cosmology?*

HB: You'd have to hide your work in a children's book or something, just like when you were back in the New Orleans library looking at science books.

RK: That's right. I'm not sure when things really turned around— maybe it was with the COBE observations; but 15, certainly 20 years later, people were saying, *"How can we hire cosmologists? What can we do to get them?"*

It just staggers me, how quickly this Standard Model of cosmology developed. Sometimes, I just shake my head. And now of course the graduate students and postdocs and young faculty members know nothing else—that's the world they grew up in.

HB: Well, kids today: they know nothing.

RK: They don't know what it was like. We had to build models in the snow.

HB: And walk 17 miles to campus each day.

RK: Right: uphill both ways.

HB: OK, so let's talk a little bit about dark matter now. Perhaps you can start off with the historical origins of the problem, then move to where we are today, before giving me your sense of what's going to happen in the future.

RK: Well, going back to the idea of a standard cosmological model, this idea has to be tempered by the fact that 95% of the universe is dark matter and dark energy that we don't understand.

In some sense, it's thrilling that we have this Standard Model that can do so much and looks so powerful, yet it's scary that there's a big gigantic hole in there representing dark matter and dark energy that we don't understand.

So let me first talk about dark matter. The idea of dark matter is not new. It was something that was pointed out by astronomers as early as the 1930s.

HB: This Zwicky guy, right?

RK: Yes. Fritz Zwicky. He was at Caltech when he did most of this work, but he was Swiss. In fact, he was the sort of a person who caused a lot of mischief, a lot of trouble. He was not the most lovable, cuddly person.

HB: Well, he was Swiss.

RK: Right: he had all the lovable warmth you would expect of someone who builds cuckoo clocks, maybe, I don't know. Let's just say that he wasn't exactly Heidi's grandfather.

HB: There goes our Swiss audience for Ideas Roadshow, by the way. Boom: completely gone.

RK: And while Zwicky was a bit of a troublemaker, when he was a student at ETH in Zurich, his next-door neighbour caused even more trouble: Vladimir Lenin. He lived for a time in Switzerland before he went off to ferment revolution in Russia.

Anyway what these astronomers in the 30's were doing was measuring the mass out there in the galaxies around us. They did this by measuring velocities.

So we can determine the mass of the sun by measuring the velocity of the earth above the sun. The sun produces a gravitational potential that causes the earth to move.

And what they found was that something was causing a gravitational potential that was causing galaxies to spin faster than you might imagine. So astronomers looked through telescopes, measured the rotation velocity of galaxies, and asked themselves, *"How much mass do you have to have in order for the galaxy to spin that fast and not fly apart?"*

And it turns out that the answer to that question is much larger than the amount of stars, or anything else, that they see. Something's missing.

This is what we call dark matter. And this was discussed and mentioned by astronomers for decades; and I don't believe physicists paid any attention to it. Georges Lemaître, one of the fathers of the Big Bang theory, wrote about dark matter—I think he used the term "dark matter"—but didn't say much about it.

There was something out there that was sort of flashing, *This is important*, but no one paid attention to it.

And then in the 1970s, there was a group of astronomers—Vera Ruben is a person whose name is often associated with this—who just made better and better observations to the point where the issue became simply unavoidable. It became increasingly difficult to just say, *"Oh, astronomers will figure it out eventually; it's not really something interesting."* It finally shook people up to the point where they started to say, *"There must be something there".*

So my PhD thesis was about dark matter: the idea of neutrinos as dark matter. These were not your basic little neutrinos that are part of the Standard Model of particle physics, but imagine you have a neutrino that has a mass of a million electron-volts or a billion electron-volts: that could be dark matter. And people like

Steve Weinberg and Ben Lee were working on similar ideas at the same time.

And that, for me, was really my entry into cosmology: worrying about how neutrinos might be the dark matter.

Over the years, people have spent increasing amounts of time on the idea that the Standard Model of particle physics may not be complete: there has to be something beyond the Standard Model that could explain the dark matter that we don't see, some kind of particle that interacts only very feebly with the normal particles—a new type of particle.

In other words, this is physics *beyond* the Standard Model of particle physics; and that's attracted more and more attention. And it became apparent that there were experiments that you could do to *test* this idea.

So if the particle was produced in the primordial soup of the early universe, then if we can somehow reproduce the primordial soup, we may be able to produce it and detect it here.

And the place where we produce primordial soup is in high energy collisions at accelerators, like Fermilab or CERN. They make primordial soup, perhaps they're making these things we call WIMPs (weakly interacting massive particles).

And personally I love this idea of an elementary particle to explain dark matter, which holds galaxies and other large scale structures together, because my original loves when I was a 10-year-old boy were astronomy, the big things, and particle physics, the little things.

I had never thought—nobody, I suppose, had thought—that there might be a deep and profound connection between the two. So it's wonderful that I just came along at the right time.

HB: OK, so two quick things. My understanding is that there's the mystery of dark matter related to rotation curves of galaxies from Zwicky and Ruben and so forth, which might conceivably be able to be dealt with by other types of explanations, such as Modified Newtonian Dynamics (MOND), but there are now other reasons to believe in the existence of dark matter, despite the fact that no direct experimental confirmations of an associated particle have yet been found. Maybe I can ask you to say a word or two about both MOND, and the other reasons why most people believe that dark matter exists.

RK: Well, it's very seductive to imagine that there's one explanation that describes many phenomena on many scales; and as you say, dark matter is seen on many scales: our local stellar neighbourhood, our galaxy, small dwarf galaxies, clusters of galaxies, groups of galaxies, up to the largest-scale structure that we see from the microwave background and how this large-scale structure is formed.

It's conceivable and possible that there is simply **one** particle to explain all of these issues, and that's naturally a very seductive idea. Of course, it may not be right. We have examples in science where we thought one thing could explain many different phenomena and it turns out not to be true.

But until there's evidence otherwise, I am inclined to believe that the dark matter is *not* modified gravity—MOND, or Modified Newtonian Dynamics—that is, it is some yet to be discovered elementary particle.

It's also very seductive to think about that because you can think of ways that this idea can be *tested* in the laboratory, by producing them at high energy colliders, like CERN.

Or perhaps they have some feeble interaction that, if you did a sensitive enough experiment and remove all the background, you could actually directly detect the relic dark matter around us.

Or perhaps this dark matter is accumulating, accreting into the centre of our galaxy, or the centres of other galaxies, and annihilating, producing some signal that we can see today.

So there are many ways to test this; and this has been going on—of pushing this frontier, developing these different tests—for probably 35 years or so. And finally we're at the position where we should be seeing something.

HB: OK, what do you think is going to happen? Give me a fearless prediction for the next 10 years, because as you say, there are a lot of additional experiments that are starting to come online now. What do you think is actually going to happen?

RK: Well, I've been consistently saying that within five years we'll have evidence. I've been saying that for 30 years now, so my consistency is something to be admired, if nothing else.

HB: OK, so clearly you are a tenacious fellow.

RK: But I actually am changing. Evolving. What is the phrase? *"My thoughts on this matter are evolving"*.

HB: That's very political. You can clearly see that you have experience as a dean.

RK: Right, right. My ideas are evolving. The jury's not in yet, but I must admit that I'm surprised at what's happened—or not happened—at the LHC at CERN. They found the Higgs boson—I wasn't surprised by that—but they haven't seen anything else. In particular, they haven't seen any evidence for supersymmetry, which is something that a great many people expected to be seen.

As far as the high-energy physics community goes, it's more a surprise that they haven't seen it than if they had.

HB: So they're about to turn the LHC back on again.

RK: Right. They're about to turn the LHC back on again at roughly twice the energy.

HB: And what are your fearless predictions?

RK: Well, I'm afraid that two years from now there will still be nothing seen and that we will abandon the idea that the dark matter is a weakly interacting massive particle, or a WIMP.

It doesn't mean that it's not a particle, but it means it doesn't really have its origin in the primordial soup produced by thermal collisions in the early universe. And if the LHC doesn't see any sign of new physics in two years—not necessarily discover a WIMP, but some sort of new physics beyond the Standard Model of particle physics—then I think it could be a scale-tipping moment and people will really start looking at *other* ideas for dark matter.

And there are many other ideas. Maybe even MOND.

HB: Just so that I understand you correctly: because my recollection of what you just said was, "*I'm afraid that if they don't see evidence of dark matter at the LHC within two years ...*"

RK: Well, not just evidence of dark matter, but evidence of some new physics.

HB: Right. Like supersymmetry.

RK: That's right. Beyond the Standard Model physics.

HB: OK. But my question was really about what you believe. I understand that you would *like* to see something new and interesting come out of CERN and other places. My question was really about what you *think* is going to happen.

RK: Okay, so I'll look at the camera and say, "*The LHC will not see anything beyond the Standard Model*". And the idea that dark matter is this weakly interacting massive particle will fall from favour; or the models, the ideas of how it can be produced, will become so baroque and convoluted that people will say, "*There must be a simpler explanation*".

HB: Do you realize that you're only the second person I've filmed who's stopped and looked directly at the camera. The other was a guy who works on the global history of Islam (UCLA's Nile Green).

RK: Well, making a prediction. I'm going to have to own it.

HB: Well, I think that's all good. I realize that's not your day job, though.

RK: Right: we're not economists or anything like that.

HB: Well, economists don't make predictions either: they make retrodictions, trying to predict the past—and often badly too. But that's another subject entirely.

Questions for Discussion:

1. Do you think there are significant outstanding issues in physics today that are analogous to the state dark matter was in back in the 1960s?

2. Why do you think most physicists are not in favour of MOND? Are there objective reasons to be sceptical of that framework? (Readers interested in this issue are referred to Chapters 7–8 of **Cosmological Conundrums** *with cosmologist Justin Khoury and Chapter 6 of* **Astrophysical Wonders** *with astrophysicist Scott Tremaine.)*

IV. Dark Energy

Particularly hard to swallow

HB: So that's very helpful, thank you for that. I'd like to move to dark energy now, asking you to talk a little bit about our current understanding of things there before asking you to give your personal views on what you think this thing actually is.

RK: Dark energy, or the phenomenon that's attributed to dark energy, is something that we see only on one scale. Let's imagine the simplest possibility, that dark energy is Einstein's cosmological constant—the idea that every cubic inch of space has a structure, has some mass density associated with it. It's very, very small, but there's a lot of cubic inches of space out there. And this affects the expansion rate of the universe.

Now, I think that modern cosmology began—although it may not have been appreciated at the time—with the 1929 discovery of the expansion of the universe by Edwin Hubble, and what Hubble did was to measure the expansion velocity, the expansion rate, of the universe today.

And until 1998, that's sort of the only information we had: the expansion rate of the universe today. Then, starting in 1998, with the work of two Nobel Prize-winning groups that discovered the acceleration of the universe, we were able to look out in space back in time and deduce the expansion rate of the universe in the past.

The expectation that everyone had—although some people now say, "*Oh, no, no, I knew it all along*"; but the expectation that everyone would *admit* to—is that the expansion rate today is slower than the expansion velocity in the past. That's because as the universe expands, you would expect gravity of all the massive objects to pull things back together.

HB: Right: gravity acting as a sort of break on this universal expansion.

RK: That's right. But that's *not* what was discovered. What they found was that there was an **acceleration** of the universe; and whatever is producing this apparent acceleration of the universe, we call "dark energy"—and again, the simplest thing to account for it theoretically is that there's some energy density associated with empty space in the form of the so-called cosmological constant.

Now, I have to admit to my own prejudice here. Scientists don't like to admit to prejudice, but in fact we *do* have scientific prejudices. Some ideas to me are like the fingernails on the chalkboard.

Dark matter doesn't really bother me. I don't lose sleep over it. My attitude is, *Oh, this is a great opportunity to use this idea. Maybe it's this, maybe it's that.*

Dark energy, on the other hand, to me is like fingernails on the chalkboard. It just drives me nuts. I don't like it. I don't like it; I admit that it's a prejudice, but there it is.

HB: So why? Why is dark energy so much harder for you to swallow than dark matter?

RK: I don't have a good explanation. It's not a logical thing. It's not that I say the observations are wrong—although I did for at least a couple of years. I kept saying, *"There must be some other effect responsible for these observations"*, until I was finally convinced otherwise.

I just can't swallow it. I think we all have these prejudices, these ideas, that we just can't swallow. We mentioned Paul Steinhardt earlier—he can no longer swallow the idea of inflation as he invented it. Some people can't swallow the idea of superstrings. It's an aesthetic thing; it's a matter of taste.

I believe that there's something very different going on, or a different interpretation of the observations. Now, one example might be a modification of general relativity. It could be that Einstein did not have the last word on gravity. That would not be surprising, but what would be surprising is if we see the first indication that there's something beyond Einstein's theory of gravity on the very largest length scales. We would have expected it on the very small scale.

So it's sort of the opposite of what you would have expected. But in hindsight, 20 years from now, when we understand what's beyond Einstein's theory of gravity, people might say, "*Oh, these people like Rocky who started this field a long time ago, they got old and they couldn't accept this new idea*". That may happen. Or it may be that it's something different.

To me, dark matter is understandable by a particle or something like that. I don't think it's going to really, fundamentally change our approach and force us to say, "*Wow, we should have been thinking about something else*".

But dark energy might well be different. If there is a revolution, if there is a real break in the Standard Model of cosmology, it will come because of dark energy.

HB: I'd like to explore your sense of discomfort a bit more. At some level, couldn't one put it this way: You're somebody who puts a very strong emphasis on the beauty and the evocative power of general relativity—

RK: It's a beautiful, complete—not quantum, but nonetheless—it's a beautiful, complete, conceptual, simple idea. Why screw around with it?

HB: Right. And, moreover, it's based upon deep, physical principles: the equivalence principle. And then you've got the Standard Model of particle physics, which is also based upon some deep principles: the gauge principle, symmetries, conservation laws and so forth. Let's forget about issues like quantum gravity and all that for the time being.

So when it comes to dark matter, you say to yourself, "*Okay, there's this extra stuff around—maybe it interacts weakly, but it clumps together: it's gravitational-like. It doesn't really mess with the general theory of relativity. I don't really know all the mechanisms to it, maybe there's some weird particle stuff that's going on. It's complicated. But whatever it is, it doesn't really bother the general theory of relativity.*"

Whereas dark energy seems like a horse of a very different colour, in that although you can easily say, "*Well, it's just the cosmological constant*", and you can throw in whatever value you want for that—although as an aside, it would also be nice if you could make that correspond *somehow* with—

RK: With what's **observed**, right.

HB: But nonetheless that *does* seem to fly in the face of the ethos and the power of general relativity. Would that be a fair way to describe your anxiety?

RK: Yes, yes. That is well put. General relativity is such a beautiful structure. The Standard Model of particle physics also has a beauty and structure of its own, but there's nothing like the equivalence principle that tells us the details of the structure: in general relativity you state the equivalence principle, and then it only took Einstein eight years or nine years to work out the implications of it.

So that's a good summary. Again, I think it's okay to have prejudices if you recognize them.

HB: Right. And it's worth emphasizing, I think, that you're *not* saying, "*I'm not going to go to this seminar because I disagree with this guy*", or "*I'm not going to read a paper that has this particular view*".

You're saying, *"This is my view, this is my bias, and until such a time as I have reason to believe otherwise, I will cling onto this position"*. I don't want to seem overly agreeable, but it seems to me that that's a completely reasonable way to proceed. I mean, you're a human being, after all. You have to have some sort of biases.

RK: That's right.

Questions for Discussion:

1. To what extent is there a correspondence between the aesthetic appeal of a theory and its objective truth-value? Does the belief in such a correspondence involve the adoption of another "meta-philosophical" position? If so, is there any historical evidence for such a position?

2. What is the so-called "cosmological constant problem" that is obliquely referred to in this chapter? (Readers particularly interested in this issue are referred to Chapter 10 of **Cosmological Conundrums** *with Justin Khoury and Chapter 3 of* **The Cyclic Universe** *with Roger Penrose.)*

V. Motivational Insights

The importance of passion

HB: I'd like to talk about something a little bit different now. You've written a popular book, *Blind Watchers Of The Sky*, that was very well received—and yet you haven't written another book after the better part of two decades. This seems to go very much against the standard physics popularization idiom. I would've thought some agent would've got their hooks into you, saying to themselves, *"This Rocky Kolb guy, he's got a good sense of humour and he can write well—he should be writing thirty more of these kinds of books"*.

RK: Well, people have asked me about it—I have been approached. I'll tell you a little bit about how the book started. I was teaching a course at the University of Chicago—many books come from courses that people teach—that I had developed over a series of years. I wanted to teach not only our present model of cosmology, but how we came to develop ideas about cosmology. What is the history of the development?

This was a course for non-science majors, people who are not so much driven by curiosity about science, but more by their curiosity about people. So I thought that if you can tell the story of the people, you can teach the science a little bit to tell the story— so when they weren't looking I would teach them some physics.

I had been teaching this course annually for about six years or so, and there's a saying: the first year you teach a course, *you* learn the material; the second time you teach the course, ***the students*** learn the material; the third time you teach the course, ***nobody*** learns anything.

So I was sort of worn out teaching this, and finally the chair told me, "*Okay, you don't have to teach this for a couple of years*".

And I said, "*I'm so relieved*". And then I turned around and wrote the book that was based on the course. I don't know why I did it. I just had to get it out of me. And I guess I had planned that it would be published, but I didn't approach a publisher with a proposal.

HB: That's very strange. Have you done other things like this in your life?

RK: I don't think so, but within a month I had finished the book—I couldn't stop it. Some years earlier I had written a textbook with Michael Turner (*The Early Universe*), and that was sort of a struggle. Maybe working with two people is harder to do than working with one—

HB: Well, it was also a very different sort of book.

RK: Yes, it is different: it's very technical, and we were writing the papers as we were writing the book, so the book was evolving as our knowledge changed.

But I found writing *Blind Watchers Of The Sky* a great joy to do, and I think I'd like to do something like that again. I don't remember how I had the time to do it, though.

HB: Have you ever gone back to teaching the course after having written the book?

RK: Yes, I have.

HB: And I'm guessing that you use the book in class?

RK: Yes, I use the book, but I don't really follow the book too much.

HB: And have you thought about writing another book? Do you have any other ideas?

RK: Well, in fact, I have a few ideas. The book ends with a bit of modern cosmology, just very superficially; and I've thought about writing a book which is more personal: my experience as being part of this wave of starting a new field, in some sense, talking about the people, and things like that.

I think eventually I will do that. Not so much about what I've done in particular, but more about the atmosphere, the people around me, the scene: it was such an exciting time.

HB: And still is.

RK: That's right: it still is, it still is.

But of course it was a different situation then. Back in 1983, Leon Letterman, who was director of Fermilab at the time, hired me and Michael Turner to start a theoretical astrophysics group at a national laboratory. This was really bringing particle physics and cosmology together. I was 32 years old, and Mike's older—he was 35 or something like that—and it was a strange sort of experience, being 32 years old being given the keys to the kingdom. He told us, "*Okay, hire some postdocs. Get some students. Hire some staff members. Go form this group.*"

And Mike and I sort of looked at each other and thought, "*Is he crazy? He's trusting **us**?*"

It was this incredibly exciting time of building something that really hadn't been built before: the people coming through, so much excitement, witnessing a new field being born.

I'm very proud to have played some role in it, and—again, in the Will Fowler tradition— of fostering others, creating a stimulating and supporting environment, and appreciating what my students and postdocs have done. Many of them have gone on to do many more wonderful things; and that's the thing I'm most proud of.

So I'll probably write a book about that. Now, my colleagues probably have a different memory of exactly how things happened.

HB: Well, the hell with them. Let them write their own books.

RK: Well, it's not going to be a "tell-all" because there's not much to tell, but I'm sure I'll get a lot of corrections. And that's perfectly reasonable. Scientists' memories are unreliable.

HB: Well, everyone's memories are unreliable, but it's important, I think, to share your unique experiences, so I would certainly urge you to write that. It sounds like you're moving in that direction anyway.

RK: It's the next book I'll write.

HB: Very good. I have one final request, and that involves a question more about pedagogy. You've talked frequently during this conversation about the importance of supervisors, mentors, and teachers.

Would you have any specific recommendations to teachers and educators—not only at a university level, but at a high-school level and perhaps even below—in terms of what they might do differently or better to recommend their students to fulfill their potential?

RK: If I could sum it up in one word it would be "passion": Have a true passion for what you do and convey that to the students, trying to ignite a passion in them. Don't be so preoccupied with the specifics of the curriculum and all these other things: it's the passion that matters most. If your student doesn't learn everything in the first 12 chapters of the textbook, that's all right. Focus on triggering the passion.

I give a lot of lectures to high-school teachers, and a few years ago, I was having dinner with about 25 or 30 high-school science teachers and sitting next to me was a recent graduate in the biological sciences who was teaching biology.

And I asked her over dinner, "*So, what excites you about biology?*" And she told some story about slime mold or something—some topic that didn't particularly interest me in the slightest. But you could see the passion in her eyes, the enthusiasm. It was beautiful. And I asked her, "*So what do your students think about this?*"

And she said, "*Oh, I don't teach that. It's not in the curriculum*".

That struck me as a terrible waste, a real lost opportunity. It's important for students to see that passion and enthusiasm in both themselves and others.

So, focus on the passion. Nothing else matters. 10 or 15 years from now it won't matter if they can reconstruct a ladder leaning against a wall, or a system of pulleys, or whatever.

HB: It may not even matter 10 minutes from now.

RK: That's right. But an appreciation, a passion, will matter. So I would urge all teachers, all educators, to focus on that.

HB: Excellent. Anything I missed? Anything you want to add?

RK: I don't think so. Thank you. It's been a very wide-ranging conversation. I enjoyed it.

HB: Thanks a lot, Rocky. That was great.

RK: Thank you.

Questions for Discussion:

1. Would you be interested in reading a book on Rocky's experiences at Fermilab? Should scientists be more generally encouraged to write first-hand accounts of the history of science and the structure of scientific development?

2. Should professional scientists be more involved in high-school science? If so, how, exactly, should that be done?

3. In what ways has science teaching changed in the last 30 years? Do you think it is improving, generally speaking? How might it be further improved still?

Continuing the Conversation

Additional Ideas Roadshow conversations not offered in this collection that the reader might enjoy include *The Pull of the Stars* with Imperial College cosmologist **Claudia de Rham**, *The Physics of Banjos* with Caltech Nobel Laureate **David Politzer** and *Examining Time* with Perimeter Institute theoretical physicist **Lee Smolin**.

Inflated Expectations:
A Cosmological Tale

A conversation with Paul Steinhardt

Introduction

Not Even Wrong

Physicists are pretty good at coming up with memorable phrases to express their scientific disdain.

Einstein famously decreed, *God does not play dice with the universe*, as his justification for denying the inherently statistical nature of the world that quantum mechanics seemed to present.

Of course, the development of quantum mechanics was known to wreak considerable intellectual havoc among even its most significant contributors, such as Einstein. Erwin Schrödinger became so uncomfortable with the implications of his celebrated equation that he even came up with a notorious thought experiment involving a half-dead and half-alive cat to demonstrate the palpable absurdity of a standard interpretation of the theory, while plaintively summing up his view on quantum theory later on in his life with a pithy, *I don't like it, and I'm sorry I ever had anything to do with it.*

Not all such dismissive remarks were specifically geared towards quantum mechanics, however. Perhaps the most notorious physics put-down is attributed to Wolfgang Pauli, who was said to have summarily rejected the work of a young physicist that was put before him by archly declaring, *It is not even wrong.*

Pauli's meaning, it seems reasonable to conclude, is that the young physicist's work was not only incorrect, it couldn't even be coherently expressed in such a way as to be clearly and explicitly falsified. For falsification, too, is a form of scientific progress—

albeit of a much less triumphant sort—that sometimes paves the way for deeper and more accurate theories.

Enter Paul Steinhardt, the Albert Einstein Professor and Director of the Center for Theoretical Science at Princeton University. Paul is a remarkably broad theoretical physicist who has made singular contributions to both cosmology and condensed matter physics. Among cosmologists he is perhaps best known for his seminal work in the early 1980s, together with Alan Guth and Andrei Linde, that led to the establishment of the theory of cosmic inflation as the primary paradigm of modern cosmology.

These days, however, Paul has decidedly broken ranks with his erstwhile inflationary colleagues, consistently drawing attention to the fact that there are deeply unsettling aspects of the theoretical framework of inflationary cosmology that should give all of us serious pause.

> *"Instead of driving the universe the way we had hoped from some random, initial state into a common, final condition consistent with what we observe, in fact the story of inflation is the following: it's very hard to start; and if you **do** manage to start it, it produces a mess—what we call a "multiverse"—consisting of an infinitude of patches of possible, cosmic outcomes."*

Given the propensity of today's theorists to invoke the notion of infinity, the reader might well be excused for not appreciating the seriousness of this problem. But this, Paul explains, is hardly the sort of thing to be swept under the rug, as it implies that, since any outcome is possible, there is no conceivable way we might one day be able to rule out the theory, even *in principle*.

Even worse still, however, sweeping it under the rug is precisely what many of his colleagues seem very much determined to do.

> *"I've had this discussion where I'll say, '**Well, what do you think about the multiverse problem?**' and they reply, '**I don't think about it.**'*

*"So I'll say, 'Well, how can you **not** think about it? You're doing all these calculations and you're saying there's some prediction of an inflationary model, but your model produces a multiverse—so it doesn't, in fact, produce the prediction you said: it actually produces that one, together with an **infinite** number of other possibilities, and you can't tell me which one's more probable."*

*"And they'll just reply, '**Well, I don't like to think about the multiverse. I don't believe it's true.**'*

*"So I'll say, 'Well, what do you **mean**, exactly? Which **part** of it don't you believe is true? Because the inputs, the calculations you're using—those of general relativity, quantum mechanics and quantum field theory—are the very same things you're using to get the part of the story you wanted, so you're going to have to explain to me how, suddenly, other implications of that very same physics can be excluded. Are you changing general relativity? No. Are you changing quantum mechanics? No. Are you changing quantum field theory? No. So why do you have a right to say that you'd just exclude thinking about it?'*

"But that's what happens, unfortunately. There's a real sense of denial going on."

Dogmatic denial is not terribly good for science. In fact, it could well be regarded as precisely the sort of closed-minded, unreflective attitude that the modern scientific temperament has emphatically, and so successfully, struggled against for centuries.

But it's not just a question of scientific stubbornness. Because theorists unwilling to grapple with inflation's "multiverse problem" aren't simply resolutely clinging to their theory despite any objective observed support for it. As Paul points out, they are clinging to their theory *independent of* any observed support for it.

Take the case of the reported findings of the BICEP2 experiment.

In March 2014 it was announced that BICEP2 had found a signal that was declared to have resulted from primordial gravitational waves consistent with inflationary theory, leading many to immediately crow that this result was irrefutable proof of inflationary theory.

A few months later, however, the signal was clearly identified with something quite different (dust within our own galaxy). Despite this observational about-face, however, proponents of inflation remain strikingly—and depressingly—undeterred.

> "So you would think, If you just declared **victory** on the basis of the discovery of them, doesn't that mean that you have to declare **defeat** on the fact that you **didn't** see them?
>
> "And the immediate response of the proponents of inflation was, '**Absolutely not. Our theory is flexible enough that we can do that too. And we immediately got a litany of papers saying, Here's how we'll do that**'.
>
> "So you might well ask, Is there anything you could observe that would tell you that inflation is wrong?
>
> "And, again, for many of the leading proponents of the field, the way they answer that question is to say, '**No. Inflation is so flexible that no test or combination of tests can possibly disprove it**'.
>
> "In fact, in their view it has three degrees of flexibility: there are initial conditions that I'm allowed to fiddle with, parameters that I'm allowed to fiddle with, and then a multiverse that I'm allowed to fiddle with. So according to them it's super-flexible.
>
> "And my reaction to that is, '**Okay, then, doesn't that mean that you concede?**'
>
> "And their response is, '**No, what's wrong with that?**'"

What's wrong, Wolfgang Pauli would dismissively inform us, is that it's not even wrong.

The Conversation

I. Scientific Beginnings

Marie Curie, Richard Feynman, and finding one's path

HB: Before we start talking about cosmology, I thought we'd take a step back and begin with your youth and how you got into science. Were you always keen on science?

PS: I was always interested in science, yes.

HB: You grew up in Miami, right?

PS: Well, my father was a lawyer in the Air Force—a judge advocate —and so we moved all over the place. I was born in Washington D.C. and we moved, when I was very small, to Alabama, for a brief period. Then, we moved to Paris, so I actually spoke French before I spoke English, but I've long forgotten all my French. I came back from France in time to start elementary school.

My father used to tell me stories and fairy tales, but part of his stories—for some reason, which I don't quite understand or know the answer to—were stories about scientists making discoveries. I always loved that. Those, to me, were the best stories.

HB: Do you remember any of them?

PS: Yes. My favourite one was about Madame Curie discovering radium. He had an elaborate story about the discovery and how they had to work so hard to find these little traces of material that they were looking for.

The exciting moment to me was always the moment when they knew something that no one else knew. I just thought that was so cool: that you could be the first person to know something. I think that's what got me interested in science from the very beginning.

When I was nine years old, my father passed away from Hodgkin's disease. My interests in science at that time were more related to biology and medicine. Those were the fields I really wanted to get into. It was definitely related to seeing if I could solve the problem that had killed him.

But I was also getting exposed to a lot of mathematics at the time, just because of the school that I was in. After my father died, we moved to Miami, and that's where I stayed through high school. Miami had a very progressive education system in that particular county—Dade County—and there were a lot of opportunities for me to take advanced math courses at a nearby university.

HB: Was there a particular teacher who influenced you strongly?

PS: There were lots of them. I really liked school, and I liked all my teachers. In my second year in Miami, I had a teacher, Mrs. Thornburg, who was particularly interested in math. She helped me accelerate, relative to the rest of the class, so that I could learn a lot of advanced math.

This was also the time of post-Sputnik, when the math programs were changing in schools, so there was lots of novel, educational material in math that you could get your hands on. You could get exposed to things like set theory and number theory at a fairly young age.

HB: Were you doing this independently or under her encouragement?

PS: I began under her encouragement, and then moved on to doing it independently. Then, about a year later, I was invited to do an original research project in math to present to the National Council of Teachers in Mathematics. They were having a convention in Miami and I don't know how they chose people, but they approached me and said, "*If you want to participate, try to find something you can do research on.*" So I found a mathematics project I wanted to do research on.

HB: What was it?

PS: It involved searching for prime numbers—algorithms for efficiently searching for prime numbers.

HB: You didn't crack the Riemann hypothesis along the way by any chance? Anything we should know that's been lying around in a desk drawer somewhere?

PS: No, no, it wasn't anything like that. I had read a *Scientific American* article that talked about certain algebraic expressions, certain polynomials, in which, if you substituted integers, they'd give you prime numbers for a certain duration before they'd suddenly give something non-prime.

And the question was, *Could you make longer and longer sets of polynomials that would give you longer and longer sequences of prime numbers?* So, that's what I was playing with as a child. It was something you could do. That introduced me to a lot of basic number theory, famous hypotheses about prime numbers and things like that. I also had a little project of my own in which I had found some polynomials—it had done an okay job.

HB: Were you doing this independently? Did you have other friends doing this too? Was there a group of you?

PS: No, this was just me independently. Once I got interested in it, I just wanted to figure it out. The idea that you couldn't predict what the prime numbers were was one of the first, real puzzles I came into contact with as a child. It's actually a very profound, deep, puzzle and that just grabbed me, like I think it grabs a lot of young people who are mathematically inclined.

At the same time, I was doing other things as well. I had a chemistry lab at home, and thought of myself as having a biology lab at home too. The space program was also happening at the time, and I was a big follower of that as well.

I just really loved science. One of my favourite books—I think it was actually one of the first books that I read when I first began to read—was a science book, an eclectic collection with chapters, each about what's going on now in a different field of science. Each one of them was fascinating to me.

It was like being a kid in a candy store: I just thought it was a wonderful playground for things. What I didn't realize until years later when I happened to look at it again was that this book was actually a collection of vignettes focusing on different scientists at Caltech, which is eventually where I went for my undergraduate degree.

HB: So perhaps it had a subconscious effect.

PS: Exactly. Here I thought I was making my choice independently but I had been totally pre-programmed. So yes, I always had broad interests; and whenever I could get my schoolwork done early enough, I was always doing some sort of research project beyond what was in school.

I liked to feel that I was doing research and discovering things, so I was naturally drawn to do something where I felt that I was discovering something that other people didn't know. I know I wasn't doing that, but I was simulating it.

HB: Do you keep reading about scientists? I mean, now that you're actually a highly respected, practicing scientist in many, different domains, are you still interested in the history of science or scientific biographies? Do you have time to pursue those interests?

PS: Much less so, probably because I feel that such biographies don't really capture the reality of the story. They're designed to focus on the individual, and I now think of science as much more of a complex interrelationship between individuals and ideas. The reality is much more interesting than the way science is often portrayed, which is by focusing on an individual, or a few individuals, when the reality is that it's a very complex interaction of ideas bubbling up here and there before, suddenly, something happens.

Books point to that moment as being the moment of discovery, but it wouldn't have happened without all these other interactions. The reality is just much more interesting—on the whole it's not really captured well, I think, in books.

HB: I'd like to get back to you and your work, of course, but just as a slight diversion—since you brought it up—why do you think that is? My sense is, from what you were saying, that popular histories of science tend to be more hagiographies—a worshipful notion of the scientist as this independent, great discoverer, this Einsteinian figure who sits in his room for 10 years and comes up with all these great ideas.

So why might that be? Is that part of a mythology that society somehow needs? Is it some elaborate form of self-justification to all of those who haven't discovered something—that it takes a solitary genius to do it?

Because as you've just said, the more sophisticated version of complex forms of collaboration is not only much more dynamic and interesting as a story, it also happens to be true.

PS: Well, first of all, I think it's harder to create such a story, because you have to consider many individuals coming in and out of the story for brief moments and making an important contribution along the way. That's a difficult thing to handle in a way that makes written sense.

It's also hard to research because, years after the fact, the smaller players have withdrawn, yet they were essential to the prime results. So it's certainly easier to write it as an individual, hero-worship story—and maybe, to some degree, it's easier to digest. I'm actually not sure about the digestion part—I think we can tolerate a few complicated stories about how things are discovered.

Occasionally I've run across books that do that for some discovery—like the history of the Internet, which involved lots of different types of organizations and types of people contributing at different levels.

But in general I just think it's probably hard to write those stories and maybe harder to digest, so people tend to write them another way, but I don't know that we need it.

This focus on heroes also leads to other phenomena, like prize-giving to individuals. Personally, I've always thought that prizes should be for ideas not people. And you should use the prizes to fund people who are working on the follow-up to that idea. That would be a real contribution to the science, I think, if you would award things in that way.

HB: That's a great idea. But that doesn't seem to be the way that things are going. There have been some new prizes created recently, but the prizes that I've seen anyway—maybe I haven't been paying sufficient attention—are all given to individual people. Of course, one doesn't want to be castigating people who are channelling their philanthropic interests towards scientists, but yours seems like a really great, worthwhile, approach. Have you ever talked to any of these rich guys and suggested that they develop a prize for ideas instead?

PS: I haven't. I think the closest that you get to that is, for example, the Simons Foundation, which has been accepting proposals for collaborations to work on certain, forefront issues in science. But that's more like the standard "proposal and review process," as opposed to a "prize process," which is a little bit different, and has a different flavour to it.

HB: And that's what would couple, potentially, to the media and to a general public awareness of these issues. Because people might then say, "*It's about this idea*," which would, in turn, not only hopefully direct people towards the content, but also move them away from this individual, hero-worship story.

PS: Exactly, they'd see that the reality is this complex combination of people with minor and major players involved in the development of the idea. You could then distribute the money in a way so that the major players would have some discretion of how it would be spent, but the minor players would also get to benefit. That's better for science, and it's a better way of representing what we do.

HB: Right. Even the Nobel Prizes—which are, of course, quite different, and more individual-focused—often involve some apportionment of ratios for perceived extent of contribution to the idea. Every time they give a Nobel Prize to three people, for example, it's not always awarded in three equal parts: sometimes it's a quarter, quarter, half, and so forth. So the idea of having different ratios or different apportionments is, in and of itself, something that people understand and appreciate.

PS: That's right. The idea of distributing an award in different fractions isn't itself new.

HB: So your idea would be to give a prize for an idea; and then designate, through some objective process, those who had been important for the development of the idea to be in charge of distributing the monies towards younger people who are working on follow-up approaches to that idea.

PS: Yes. You'd have the honour, of course—the honour would still be there, you'd be one among the honoured few—but instead of money you put in your bank, you'd have money that you'd put directly into the field. I think that's a better way of doing things.

HB: You should really write something about this. I would vote for you to be the Science Minister or something like that—although that would, unfortunately, remove you from doing actual science, so it's probably a bad idea. Personally speaking, these prizes have always driven me crazy. I realize that they're good for public awareness and so forth...

PS: I actually find them more bad than good. All my worst experiences were with people who were eager to get prizes, so I think the recent trend to increase the number of such things just makes it more intense and leads to some of the worst behaviours that I've seen.

HB: Interesting. So, let's get back to your own scientific development. You've long had very wide interests—presumably they're not quite as wide, scientifically, as they were when you were much younger—and, at some point, you decided that physics was going to be your thing. You mentioned going to Caltech—I'm not sure if you were studying other things there—so tell me when physics started playing a greater role in your scientific development.

PS: Well, it was basically the first weeks at Caltech that influenced me. By the time I went there I had done a lot of work in math and I had actually done some research in biology, so I was thinking I was going to do one or the other but I wasn't quite sure which.

Of all the courses that I had taken in school, my worst courses were physics courses, so I had a pretty bad impression of physics coming in to Caltech, but after a few weeks of arriving I was exposed to the Feynman Lecture Series, because everyone had to take physics.

One of my reasons for going to Caltech was that I had this vague feeling in the back of my mind that I might like physics if I gave it one more shot. I don't know why I had that thought, because I didn't really know much about it, and if I went to some place other than Caltech where I didn't have to take physics, I probably would have filled my schedule with math and biology.

But at Caltech you had to take math, physics, chemistry, biology —there were lots of science requirements—and I figured that going there would make sure that I tried all of those things and gave everything a shot.

HB: That's a strikingly mature attitude for a young person.

PS: Well, thank you. Within a few weeks we began the Feynman Lecture Series. They're not so much about physics, they're more about how to think. Suddenly, all the ways I had studied science up to that point seemed wrong, and there was a new way to actually use your mind to figure out how things work. You could actually figure it out with your own mind and not learn it from a bunch of rules. That just changed my life.

HB: And Feynman was there in person.

PS: Yes, Feynman was there. He wasn't teaching the course that year, but he was also a very strong character on campus and, while I was at Caltech, I had the good fortune of working with him in various ways.

HB: How, exactly?

PS: The first way I interacted with him was organizing a pseudo-course called "Physics X".

HB: *You* were involved in Physics X?

PS: Yes. My roommate and I put together—well, there may have been one version that existed a few years before us, but my roommate and I went to Feynman and asked him if he would be willing to put together Physics X, and, much to our delight, he agreed. That was extremely influential.

The idea of Physics X, as you probably know, is that one afternoon a week, he would come and discuss whatever science you wanted to talk about. There were various rules about what kinds of questions you were and weren't allowed to ask–

HB: Like what?

PS: He didn't want people who were erudite, he wanted people who were curious. So you couldn't come and ask him things like, *"Tell me about the such and such equation."* He wasn't there to be a dictionary. But you could come and ask him about a specific phenomenon, like, *"What colour is a shadow and why?"* Sometimes there would be something he instantaneously knew about, but more often than not it would be something that he'd be struggling with, it would be something he hadn't heard of, so we'd see him really engage with the question.

But the most important thing to me was that every question was considered interesting. There was no sense of, *This is what I do, that is not what I do.* **Every** question was interesting. That was really influential to me, because it meant that, although he was a renowned particle physicist, a very famous particle physicist, he didn't only think about particle physics. He was interested in everything.

Now, this was from the point of view of an innocent undergraduate who hadn't done much reading about Feynman's personal history and had no clear sense of how many different scientific problems he had worked on.

HB: He was, famously, remarkably broad in his scientific interests.

PS: Yes, but as an undergrad, I hadn't seen someone like that at close range. It's one thing to hear of someone like that from a distance, but this was someone actually in the room with you.

The other fun part about Physics X was that, every now and then, someone would ask a question and for some reason he would break out into one of those stories that later appeared in books like *Surely You're Joking, Mr. Feynman!* So I got to hear them from the horse's mouth, not in some fully prepared form, but in some spontaneous form—those were always fun too.

HB: Did you use to ask questions yourself in Physics X?

PS: Yes. I'd always try to come with some sort of phenomenon or some sort of thing that I was trying to figure out—it was like a homework assignment for me, to try to come up with something—but you didn't always get a chance to ask a question every week: usually two or three questions would manage to be covered in such a session, and maybe not even that, maybe only one.

HB: I had no idea that you were involved in this. I'd heard about Physics X, of course—it's an integral part of the Feynman legacy—but I'd never thought to ask who the undergraduates were that put it together. Did anyone else ever try to do something similar? Was it imitated at all by other faculty?

PS: No. It was unusual, and really stemmed from his personal interest in doing it. The other thing that was odd was that, not only did no other faculty volunteer to do something like that, but, while you might think that there would have been a packed room every time, that wasn't the case.

It was mostly a fairly small group of people. When it would first start each year we had a fairly large group, but it would pretty quickly shrink to a core group of 10 or 20. It was great for me because I got to spend more time with him, but it wasn't like the room was always packed.

HB: Why is that, do you think?

PS: I think people make themselves busy or something, I don't know. Since I wouldn't have missed it for anything, I can't answer that question. I can't imagine missing that privilege.

Later I took quantum field theory with him, and that was really interesting because you wouldn't have recognized that it was a quantum field theory course. The typical stuff you do in a quantum field theory course is Feynman diagrams and related things, but that barely slipped into the course. He wanted to talk about lots of other things.

HB: Different ways of looking at things, or just different things altogether? I remember reading something about what had apparently motivated him to come up with quantum electrodynamics, and somehow it started from him watching spinning plates in a cafeteria. I can't imagine being taught quantum field theory that way.

PS: I'll be honest and say that I didn't get that much out of the course because quantum field theory is already hard material. Nowadays it's been restructured pedagogically with certain core material and taught in a specific way, but at that time—this was the 1970s—I took quantum field theory from three different people and no two courses were the same. Feynman's content had a lot of solid-state physics in it. Today that wouldn't seem so strange but at that time, it seemed stranger.

HB: That was a sign of the way he thought, right?

PS: Yes, and probably because of his interests at the time, he was a lot more focused on that, and the particle physics barely got into it.

Anyway, I also got involved with Feynman in a third way. I did my senior research project with him.

We worked on two problems—one was a problem I had brought, which was about Super Balls. A Super Ball is a rubbery ball toy that bounces very high, but when you bounce it around corners or under tables, it does some weird things, and eventually comes back to you. And the question was, *Why?* That was my first project, working out the way Super Balls bounce.

HB: How does that work?

PS: Well, by "work it out," what I mean is that the key physics is that it's not just elastic but it also doesn't slip, so it's close to a perfect roll. You can't slide it along the table, it rolls along the table; and it also means that, when it bounces, if it has a spin, it has to reverse its spin, so it's reversing its angular momentum as well as its linear momentum. That's the thing that gives it its seemingly peculiar characteristics.

So I worked out a set of matrices that would describe an incoming momentum and a surface and what its output was going to be. Then, once you have that matrix, if you have a combination of surfaces, you just have to multiply the matrices together to work out what the final trajectory was going to be. But for me, it was just a fun project. The fun part was that you could solve this problem entirely yourself.

The other project we worked on followed from a colloquium we had attended about something that's called a solitary wave. Nowadays we call it a soliton. It's just a wave that has the property that, even though it's a non-linear system, it will maintain its form if it collides with another such wave: it will just pass through it. Anyway, we went to this colloquium and we talked a little bit after it, and Feynman was very sceptical that this was correct.

The talk was given by some engineer—I don't remember who. So I said, "*Okay, well, we can go on a computer and we can try to work out the equations for it and see if it really happens.*" And it does, the engineer was correct. But that was a project that was essentially stimulated by a talk we had both seen together.

HB: My understanding is that Feynman had a very strong interest in undergraduate teaching and not a very strong interest in graduate teaching, which is pretty well the opposite of many other professors, many of whom resist teaching undergraduates as much as possible. Was that right?

PS: Well, he certainly stimulated the undergraduates. I didn't have that much contact with the graduate students to know what the influence was within that group. At that time, both Feynman and

Gell-Mann were there, and Gell-Mann was probably more focused on graduate students and postdocs. Feynman certainly had broad interests, and I think he was just interested in interacting with many different people, both undergraduates and the general public—he gave a number of public talks and things like that.

HB: Okay, so getting back to your career, you've had this formative experience at Caltech and presumably at that point you were sold on the merits of physics and physics research.

PS: I didn't know what physics I wanted to do, so I decided to spend every summer doing a different form of physics, and I would then decide which one I would choose.

   If I think about what happened afterwards, I thought I made a choice at the end of that, which was to do particle physics. But one summer was spent doing general relativity, one summer was spent doing condensed matter theory, one summer was spent doing particle experiments. But none of them have really gone away. Although as a graduate student I focused on particle physics for a few years, by the time I finished graduate school I was really eager to try some other things, and all of those other things came back in unexpected ways.

HB: You strike me as intellectually broader than most theoretical physicists. Would you agree with that characterization?

PS: I think that's fair. I'm not alone, though. I know a number of colleagues at Princeton who have worked in several different areas that are quite different. I think there are different types of scientific minds.

   One way I think about it is that some people come to science and end up finding a particular burning question that they want to answer, like, *I want to understand the fundamental constituents in nature*, or *I want to understand the fundamental forces*.

   Then there are others who are in my category, who are looking for a really good puzzle where they can discover something. I want to discover something new—I don't actually care that much what the field is, but I want it to be an exciting discovery in something new.

Those are two very different ways of searching through possible projects to work on, so it took me a while to decide that I was really going to commit myself to that because there's a lot of pressures to be the first kind: all the social pressures are to be narrow, and all the financial pressures—all the granting pressures and so forth—lead one to be narrow as well.

But at some point, I decided, *To heck with that, I'm just going to try it my way, and if that doesn't work then I probably chose the wrong thing to do.*

HB: Not to beat this to death, but as you're talking, I'm thinking that that's a very Feynmanesque sentiment.

PS: I am very much influenced by him, yes.

HB: My question, then, is, Do you think he influenced the way you looked at things, or do you think the reason you resonated so strongly with him is because you had those predilections right from the beginning—it wasn't so much that your character was formed by encountering him, but there was simply a natural fit between him and who you were?

PS: I think it was both, to be honest. There was a natural resonance because I already had the broad interests, but it, sort of, reawakened that in me. You know, as a student, you have this view that, *I have to figure out what to choose.* But suddenly, with Feynman, I realized, *Oh, you **don't** have to choose. Physics can be applied to **anything**.*

That was the really powerful message for me: if you have the mind of a physicist, there's no subject that you can't explore, and if you discover something interesting, it doesn't make a difference whether it's about a piece of dust or a star. It can still be really important and really interesting.

HB: And of course, he also had a strong interest in biology, which he maintained throughout his life. Have you maintained some of your broader interests in science, writ large, in the biological sciences or chemistry or geophysics or anything like that?

PS: Well, I guess I'd say that, if you're the second type of scientist where you're looking for something to work on that could be an interesting discovery, you're always open to new ideas. I think of myself as being an intellectual predator. I'm always having my ears open and quietly thinking about various things in broad sets of fields. Most of those things never see the light of day because they don't get anywhere. It's all science: there are no boundaries to the kinds of science that I'm interested in. But you only hear about the things that turn out to work.

Questions for Discussion:

1. Do you agree with Paul that scientific prizes awarded to individuals are often "more bad than good"?

2. To what extent is our educational system responsible for instilling the view in students that they "have to choose" a certain narrow area of specialization? How, concretely, might we take steps to counteract such a notion?

3. Which of the two types of "scientific minds" that Paul describes in this chapter do you most strongly identify with?

II. Inflationary Excitement

A captivating talk and subsequent insights

HB: Tell me about how you first got interested in the ideas of cosmic inflation. You were one of the people who were involved in the development of the theory. How did you get exposed to those ideas?

PS: Well, first of all, after explaining to you that I had broad interests, I have to confess that, up until I was a postdoc at Harvard, cosmology was not something that I had ever paid any attention to. In those years, cosmology didn't have a great reputation as a science—it was often described as boarder-line metaphysics, or something like that.

The situation was actually a little better than that because the cosmic microwave background (CMB) had been discovered and a lot of work was being done at the time on nuclear synthesis, but somehow that didn't filter down to me, so it was just an area that I didn't pay attention to.

When I took a course in general relativity, it happened to be taught by a mathematical physicist who didn't teach the cosmology part of it, he just taught the general relativity part of it. So I really knew nothing about it. I knew a lot about particle physics by the time I was a postdoc—that was what my PhD was in: rather abstract questions in quantum field theory and grand unified theories. I had also re-developed my interest in condensed matter physics and was beginning to work on some projects there, but I hadn't touched anything related to cosmology.

Then in 1980, at the weekly theoretical seminar in high-energy physics, I attended a talk by this fellow I had never heard of before named Alan Guth who was talking about something called "the inflationary universe", and I was just floored.

It was an amazing talk. Alan's a very good speaker, he had a really good story to tell, and he told it in a way that made it really easy for me to understand. He began with a whole introduction to cosmology: what the puzzles were in cosmology, what we knew and didn't know.

It was great for someone like me who knew nothing about this. I was listening to this talk and thinking to myself, *OK, he's told me this story. We understand a lot more about cosmology than I had realized before, but now there's this puzzle: why is the universe so homogeneous and so uniform? Why is the universe not curved when it could be?*

Then he had this great idea that this could be due to, of all things, a phase transition that occurred in the early universe. A phase transition is like the transition that occurs when liquid freezes and forms a solid, so what he was talking about was like condensed matter physics but on the scale of the universe, and I thought, *Oh, okay, **now** he's talking about something I recognize, but I had never thought about it as being relevant on a cosmic scale.*

And then this phase transition turned out to be due to a Higgs-like field associated with a quantum field theory, a grand unified theory, which was the **other** thing that I was working on. So now, the other two things I was interested in seemed to be somehow related to a third thing, which I didn't know anything about, which was cosmology.

And the whole idea seemed like a beautiful solution to the problem he had described: just by going through this phase transition, it would trigger a period of very rapid, accelerated expansion, which would smooth out any initial inhomogeneities in the universe, leaving behind a universe that was smooth and flat.

I thought the talk was over at this point because he had solved this big problem, but in the final five minutes he explained why the idea fails: the problem being that, once this inflation starts, it's so rapid, it creates volumes of space so rapidly, that it never ends.

He spoke a little bit about some work he had done with another physicist, Erick Weinberg, that began to show that there's no way of getting around this problem—and that, suddenly, was the end of the talk.

I just sat there, stunned, feeling like I had suddenly crashed. It had started off being the most exciting talk I had ever heard, and then it seemed like the most depressing one: how could such a beautiful idea like that fail?

So, I thought, *OK, I should think about this; I can give it a few weeks and think about it and see if I can find my way around it, because there must be some way around it.* Since I knew something about phase transitions, I thought, *Well, maybe there's something you can do with phase transitions that can help you get around it.*

That's how I got into it—thinking it was going to be a few weeks' digression from my regular work, but this digression has not ended since. Like a lot of things I do, this is one of those projects that had many twists and turns—it's not a simple line from how I went from there to what was the first workable model of inflation.

He had this phase transition, and the way he was going to end the phase transition, the only way he knew to end a phase transition, was by spontaneous nucleation of bubbles that would carry you past the barrier. You had an energy barrier that was keeping you in this phase that was inflating and you needed to get past it and his idea was to quantum tunnel through it, producing bubbles—and this bubble nucleation idea had actually been developed by my thesis adviser, Sidney Coleman, so I was familiar with that.

But the problem was that you couldn't produce the bubbles fast enough compared to the inflation—that's how he got stuck. So my idea was, *OK, he was assuming that the universe was perfectly uniform, but what if there were defects in it like there are in solids?* Those defects can seed nucleation events much more rapidly than if you wait for it to happen spontaneously.

And there was an obvious such seed in the story, which happened to be a magnetic monopole. We were trying to get rid of these magnetic monopoles by inflating them away—according to Guth's idea—but, another way of getting rid of them would be—well, I call it dissociation: their core simply expands and this core becomes the source of this phase transition.

That's what I worked on for a period of months. I developed that idea and showed that, while it didn't quite work, it involved a lot of interesting, new ideas. And interestingly enough, even though it didn't quite work in the cosmic setting, I thought, *Well, maybe we can do it in the laboratory.*

In the laboratory, the analogue of a monopole is what's called a "vortex," like a vortex in a superconductor. So, the question became, *Could you find dissociation of vortices in a superconductor or in a superfluid?* And I began to look for examples.

I found an interesting system—the Helium-3/Helium-4 mix—which turned out to be a nice candidate, but more importantly it also turned out that it had a type of phase transition which I had never heard of before called a spinodal decomposition.

Normally, in the kind of phase transitions that Alan had proposed, you have an energy barrier that prevents you from ending the phase transition, so you get stuck there by the energy barrier, and you just can't tunnel out.

But what happens in the spinodal transition is that, as you cool the system, the barrier just disappears, so now there's nothing to prevent you from getting out.

Once I realized that that's what the phase transition in this Helium-3/Helium-4 mix was, I had a solution that was needed for inflation: *why can't we have a phase transition like that?*

So that was the line of reasoning that led me to wanting to do something like that in the cosmos.

I had developed this idea in the summer of 1981, and that fall I moved to the University of Pennsylvania and I had a young graduate student, Andy Albrecht, who was looking for a project. I said to him, "*Let's work this out,*" and that ended up being how so-called *"new inflation"* or the first model of inflation that we thought could actually end the inflation—not just have the inflation but could actually **end** it—was born.

HB: And what was the reaction to these ideas? Was there scepticism? How did that play out?

PS: Alan was a very effective communicator and I think many people were excited by his idea—I don't think I'm the only one that might have had the reaction that I did, but it depended upon the circles in which I was speaking.

If there were people who had never heard of any of these ideas before, the reaction could be quite harsh. I won't name the place, but I remember giving a seminar very early on, one of the first seminars I gave on this idea of how to end the phase transition, and it was stopped. That was the first time I had ever heard of or seen of a talk getting stopped about two-thirds of the way through.

Although they gave me some excuse like, *"Someone has to come into the room,"* or something, I was convinced that they thought I was a complete nut by talking about this stuff. But in major institutions, people found it immediately very interesting.

But at this point it wasn't a complete idea. There were two immediate problems.

You'd smooth the universe out, but you had to figure out how to get back matter in it, which meant that you had to, somehow, reheat the universe afterwards. That was the first project that I worked on after we developed the idea.

The bigger problem, though—the really worrisome problem—was that this inflation mechanism was so effective that it seemed to make the universe *so* glassy-smooth that you'd never make structure in the universe: it would be too smooth. It was overkill: you had to figure out some way to de-smooth it.

So, what in the world was going to do that? Well, if it was going to happen during the inflation, the only thing left out of the story was the quantum physics, so the question was, *Could quantum physics solve this problem?* After all, if inflation, without quantum physics, would make it glassy-smooth, that means the energy everywhere in space would identically be the same, but quantum physics doesn't allow that to happen: it forces energy to fluctuate, it's going to produce some kinds of fluctuations.

And as you really begin to think about that problem, you realize, *Oh, this is a real threat, because we thought we made the universe smooth but now, we actually may have ruined things when you take the quantum physics into account*—as you have to do since it's there in nature. So maybe the idea doesn't work after all.

So, in the first months, there were several scary moments as we began to first do calculations of the perturbations.

This is work that I began doing with Michael Turner at the University of Chicago, and this issue of how to do this computation wasn't straightforward. Nowadays, it's straightforward and we teach it in our classes, but there were a lot of subtle questions about how, in general relativity, you properly account for the fluctuations and distinguish what is a real, physical fluctuation that might form a galaxy from an artificial, mathematical result resulting from a particular choice of non-uniform coordinates.

That was a subtle problem that people had known was a problem for, probably decades, but a good solution had been worked out by a theorist at the University of Washington named John Bardeen. So, Mike and I contacted John and said, *"We'd like to use your method for our problem, do you want to collaborate with us?"* That's how our paper ended up being the three of us collaborating.

Questions for Discussion:

1. To what extent does the phase transition Paul discovered while investigating an experimental analogue to a cosmic setting demonstrate the importance of combining theoretical and experimental approaches in fundamental physics? Do you think that most theoretical physicists today think about turning to a laboratory environment as a way of testing or clarifying their underlying theoretical ideas?

2. Are you surprised at Paul's anecdote of how one of his early seminars on cosmic inflation was abruptly cut short? Could you imagine such a thing happening in other academic disciplines as well? What do you think such a story indicates about the sociology of professional physicists?

III. Progress, Tuned Appropriately

The Nuffield Workshop and its aftermath

PS: By this time, I had mentioned this issue to Alan and he began to do his own calculation, and it turned out Stephen Hawking was doing his own approach, as was Alexei Starobinsky. And about a year or so earlier, Slava Mukhanov and Gennady Chibisov had also thought about this idea.

So it wasn't just one of us—it turned out that there was a bunch of us. The papers on inflation came out in early 1982 and then later that summer, Stephen Hawking organized a meeting called the Nuffield Workshop in Cambridge in which most of those people were gathered.

We all knew that this was the thing we had to work out, and we all knew that there was competition there. All of us came to the meeting with our own kind of different rough answer, but by the end of the three weeks everything converged.

I'm sure each of us has a different version of how this came together, but my own view is that, although different methods were giving different answers with different subtleties that had to be worked out, the nice thing about the method we were using—and that's why it's part of the "permanent story" these days—is that it was the one mathematically trustable method where you could solve this problem that I was describing that's called the gauge invariance problem: how do you separate out the real, physical fluctuations from the mathematical "coordinate" ones?

Ours was the one method that had that. The others were essentially eliding over that issue by giving heuristic arguments. Now, it turned out that the heuristic arguments that they used *did* give the right answer, but you could have easily changed those heuristic arguments so that they gave a different answer. So, *after* the fact, you could be sure they were right, but if those had been the *only* approaches used, then I think we would have been debating it for some time. But the fact that all of the groups ended up converging on the same thing was, I think, a rather important event.

HB: When you were working on this, how much interaction was there between all the different people?

PS: It was pretty intense, because we were all staying within a mile or so of one another in Cambridge. Actually, most people were staying in the same place—I wasn't because we had a baby and my wife and I were staying off by ourselves. We had to walk in sometimes to catch up with the rest of the group, but it was basically an intense session of 24 hours a day for three weeks, at the end of which it was beautiful to see that the answers agreed.

That was when the subject really took off, because no one, including Alan, had thought about the quantum fluctuations when he originally proposed the idea. And now, suddenly, there was good and bad news about these quantum fluctuations.

The good news was that they were nearly scale-invariant: that is to say if you think of the fluctuations as a combination of waves of different wavelengths, the amplitude of the waves that you were summing were more or less the same—approximately the same from wave to wave—not perfectly so, but approximately so.

That turned out to be something that theorists had speculated ten years earlier, just based on looking at the sky, might be the kind of input you need to explain the distributions of galaxies we see in the sky. So now it came out as a *consequence* of the theory, which was the good news.

The bad news was that the amplitude of those waves was way too big—too big to be consistent with what we observe. So that meant that the idea to end inflation that had replaced Alan's original idea—which was just to make the barrier go away, wasn't enough: you needed to do something else as well to, not only get the near scale invariance—that you got—but now to get the amplitude of the fluctuations to be as tiny as we knew, even then, they had to be. Later we actually measured that amplitude, but we knew from measurements at that time that it had to be below a certain bound—these had to be tiny, little waves instead of giant waves.

So although it turned out that round had failed, we were immediately able to say, *"Okay, this is what we need instead."*

We still wanted to save some of the ideas that were there before: a field that has an energy associated with it that is large and positive which drives the inflation and which is not protected by an energy barrier, but you needed to carefully fine-tune that field, in terms of the strength with which it interacts with itself and with other interactions.

In fact, you needed to extremely fine-tune it, but if you did that, then you could get everything you wanted: the smoothness, the flatness and, now, density fluctuations that could have seeded galaxy formation, that we might some day even see in the cosmic microwave background.

There was a cost, however, which was this fine-tuning. At the time the view was, *We'll solve that fine-tuning problem, we'll find out why that has to be the way it is, but it's great that we have everything else.*

So there was a tremendous amount of optimism, and a lot of focus on the fine-tuning problem, which has endured for 30 years as we've tried to address it, but never really solved it.

It looks like one of the problems of this inflationary idea that we don't know how to get around is that it has different parameters in it, different interactions of the field with itself that have to be carefully adjusted to make it work. It's not its worst problem, but it was the first major one, and the one that most people have worked on.

And by "worked on," I mean they've tried to find alternative energy forms, sources of the energy that could drive the inflation. And when you look at them, you may not spot the fine-tuning right away, but if you look carefully enough you can spot it: it's always there, hidden someplace in each of the theories. We've never gotten around it. Current attitudes about that aspect are, "*Well, maybe we have to live with fine-tuning.*"

HB: And in the meantime, the observations took off.

PS: Yes, so this inspired various sets of observations. One of the predictions of the theory is that the universe should be flat. And at the time, what we thought that meant is that, in addition to ordinary matter, there must be enough dark matter to bring the total energy density of the universe up to the critical density.

In other words, we knew that the ordinary matter that we observe every day consists of only 5% of the critical density needed to make the universe flat, so that meant that we had to find another 95% some place.

Would it be something involving very light relativistic particles like neutrinos? Or something very massive and weakly interacting? Or other possibilities? It spawned this whole industry—which, in the astronomy community, had already had somewhat of a fledgling start: they already had found some evidence for dark matter, but they didn't really have a good handle on how much. And it took about 20 years to settle that "how much?" issue. So that was one kind of industry.

The other industry was the field of closely investigating the cosmic microwave background, which was reinvigorated because now there was something definite to look for: if this idea was right, then we should see a near scale-invariant spectrum of hot and cold spots in the microwave background.

I have to say that I was a bit naive: I thought that it would take 50 years or so before we ever had to really worry about that. At the time, no one had ever measured any anisotropy in the microwave background and no one had any idea how long it would take.

Remember I'm coming from the particle physics field—I'm not interacting with people in the cosmic microwave background field, which was a small, fledgling, isolated field on its own then. But it took less than 10 years before we began to see fluctuations in the microwave background from the COBE Satellite experiment; and one of the first things that experiment declared was that they were scale-invariant.

One thing that many of the astronomers didn't absorb from that exchange of various groups associated with The Nuffield Meeting was that all of the other groups came out and declared that the spectrum should be scale-invariant.

But actually a more precise calculation—since our calculation was more rigorous, it had that precision—said, "*Actually, it's not **completely** scale-invariant; it should be slightly tilted away from scale-invariance.*" So, our paper was called "*Nearly Scale Invariant*" not "*Scale Invariant*", whereas if you look at the other ones, you'll see they talk about scale invariance.

When the COBE results came out, I realized something that they were missing was that they only fit it to scale invariant spectrum when they should have been fitting it to *nearly* scale invariant spectrum, so that's how I got into the microwave background game.

The first thing we did was we wanted to calculate the power spectrum, the theoretical predictions for models, explicitly including the effect of near scale-invariance.

The other thing that was missing was that you should, in inflation, also have a spectrum of gravitational waves that would leave an imprint on the microwave background. And that wasn't included either.

Furthermore, the gravitational waves should produce a polarization; and they didn't even discuss polarization. So there was a lot to do.

After COBE, my focus, for the next few years, turned to the microwave background and how you would look for this evidence of scale invariance tilt and gravitational waves. Now it's an industry, but I think we wrote the first computer codes to do those calculations, to point out the degree to which you can and cannot measure things.

HB: What were your conclusions?

PS: The conclusions were that you should be able to measure this tilt but you have to watch out, because at the same time that inflation produces, through quantum fluctuations, fluctuations in density, it also produces fluctuations in space-time that form gravitational waves.

They should be more or less comparable in strength because they're being produced by the same physics, but you have to watch out for that because that will produce a big effect when you begin to compare the temperature fluctuations on large scales compared to small scales: it will very much change the relative distribution that you'd expect. So if you didn't take that into account, you might conclude that things didn't fit when they actually did.

When it came to polarization, the nice thing was that was one way of separating out the gravitational wave contribution. Then you could go back and do the temperature calculation more accurately. So we outlined how to do that.

We also pointed out that there were some degrees of freedom that were more difficult to separate in the microwave background because some features—like when you look at the distribution of hot spots and cold spots on different scales—didn't just depend upon one parameter—like, say, the amount of ordinary matter— but might also depend on the amount of dark matter. Or it might depend on other features, like the tilt.

So we were trying to turn this into a program of, *What can you learn from the microwave background about different cosmic features, either individually or in combination?*

And if the best you could do was look at combinations, the next question would be, *What other data could you bring in to separate this degeneracy?*

Those were the kinds of things we were doing, which then became a big industry in the field and continues right up to the present day.

Questions for Discussion:

1. What does it mean, exactly, for a field to be "fine-tuned" and what would a "solution" to the associated "fine-tuning problem" look like in principle?

2. To what extent might a "fine-tuning problem" be "solved" by invoking the anthropic principle?

3. What do you think Paul means, exactly, when he talks about "separating this degeneracy"?

IV. Two Major Issues

The initial value problem and eternal inflation

HB: And while this was going on, your thoughts on inflation as a principle—standard inflationary cosmology—are moving in what direction? What other concerns did you have?

PS: Well, shortly after inflation was developed, two big issues arose. Fine-tuning was not one of these first big two—I'm putting that as a third issue—so although that's the one that most people spend their time on, that's not one of the real problems, in my view.

The two big problems are:

First, we didn't properly think through how inflation gets started. What we said is, "*If you have some random distribution of matter and energy coming out of The Big Bang, inflation will smooth it out.*" But we began with, "*If you have inflation...*"

Well, what does *the inflation* need? It turns out the inflation needs a universe which is rather smooth and flat to begin with, which was the very thing inflation was supposed to be doing for you.

And it needs it to be smooth and flat over a fairly large scale—larger than the size of the horizon, the largest distance that you could see at the time. So it has to occur over a scale in which normal, physical processes wouldn't be able to interact. This is what we sometimes call the "initial conditions problem": we don't know how to initiate inflation.

That is, instead of it taking over easily, inflation can only take over if someone has already smoothed out the universe to a significant degree, which solves the problem that we wanted to solve to begin with.

Suppose we say that, coming out of the Big Bang, that's very unlikely but it's not impossible. It could be by chance that it came out that way. It seems like it requires a conspiracy over large scales, so it's very, very unlikely. But it's possible. That's true. But, as unlikely as that is, it turns out that the condition you need to start inflation is exponentially more special, more unlikely.

So in order to explain your first unlikeliness, you've actually had to go to a situation which is exponentially **more** unlikely.

That was first pointed out by Roger Penrose using a very clever but subtle argument; and then, over the decades, other arguments have been developed.

HB: Wasn't it 10 to the 10^{100} or something crazy like that?

PS: Yes, I like to describe it as the worst prediction ever. You're trying to explain why we are the way we are by following this line of argument, and by using the statistical measure he proposed— which turns out to agree with other ways of doing the estimate— you end up saying, We're only likely as one part in 10 to the 10^{100}, as an upper bound.

So, that's the initial conditions problem: it's an extremely bad, disastrous situation.

Now suppose you take the point of view that some physicists today do, saying, "*But we don't understand the initial conditions, and maybe something in quantum gravity solves them*".

Well, my first reaction to that is to say, "*If you really have such a thing, then you don't actually **need** inflation, because you've just solved the problem*".

But let's suppose, somehow, that we can put that whole mess to the side. But now you run into the next problem.

So we're now assuming that we somehow managed to get inflation started, and we thought—through the ideas that Linde and Albrecht and I had introduced that I spoke about earlier—that we had figured out how to get it to end. But it turns out that we were wrong: we misunderstood the quantum physics part of the story.

When we first calculated the rate at which the phase transition that would end inflation would occur, we ignored quantum physics. Then we said, *"Oh, it made the universe too smooth, we'd better include some quantum physics to make the fluctuations."* The idea was that including the quantum physics would only have a very weak effect, only weakly perturb the story, so we would wind up with a nearly smooth, flat universe with some regions that are just slightly hotter or colder than average.

Now, what's causing those regions to be slightly hotter or colder than average? It's because when the field that controls when the phase transition ends is subjected to quantum fluctuations, this results in inflation ending a little bit earlier in some places and later in other places, leading to these little variations in the CMB—what people refer to as anisotropies.

But quantum physics doesn't let you stop there. Every now and then, there's going to be some quantum fluctuation that keeps the inflation going for much longer than you expected. It's going to be a rare event, but it's a rare event that, when it happens, produces a tremendous amount of inflation.

If you like, then, it's rare at the beginning, but if you look a moment later, it's actually most of the stuff in the universe. So you might have thought that inflation had ended almost everywhere in the universe except for a tiny little speckle over here, but the result is actually the reverse: it's a tiny little speckle where it ended, and everywhere else is still inflating.

So that leaves a chance for this to repeat itself; the part that's still inflating can, once again, end the phase transition, but there will always be a rare region that takes over again, leaving you now with two, or three, or four of these patches—and, in fact, this process occurs eternally.

This is what we call "eternal inflation." We **thought** we had ended inflation, but we had failed, because when you properly take into account the quantum physics, it is eternal. I believe I had the first model of eternal inflation, and then Alex Vilenkin came by independently with another in 1983.

And this was really a nightmare, because the second problem isn't that you just produce many patches, but the patches aren't the same. The same quantum physics that's driving this whole thing means that some of these patches, depending upon what quantum fluctuations they underwent before the patch formed, will be flat —like we observe—but some of them will not be. Some of them will have near scale-invariant spectrum like we hoped, but some of them won't have; and, similarly, some of them will be smooth and homogeneous, but some of them won't be.

In fact sometimes these patches collide, in which case you get regions that have lots of matter in them, but they're very non-uniform and anisotropic and have none of the properties we want. And because this process is eternal, you actually get an infinite number of every one of these possibilities.

So instead of driving the universe the way we had hoped from some random, initial state into a common, final condition consistent with what we observe, in fact the story of inflation is the following: it's very hard to start; and if you do manage to start it, it produces a mess—what we call a "multiverse"—consisting of an infinitude of patches of possible, cosmic outcomes.

Questions for Discussion:

1. *In what ways does this chapter illustrate how some areas of science— such as cosmology and aspects of evolutionary biology—have a particularly difficult time developing falsifiable statements? Those with a particular interest in this topic are referred to two Ideas Roadshow conversations where it is discussed in further detail:* **Science and Pseudoscience** *with Princeton historian of science Michael Gordin and* **Astrophysical Wonders** *with Institute for Advanced Study astrophysicist Scott Tremaine.*

2. *To what extent are we justified in assuming that our current laws of physics existed in the same form in the very early universe?*

V. Cosmological Denial

Resolutely clinging to a good story

HB: OK, so if I'm somebody who's watching this—maybe I'm a lawyer or a political scientist or a gardener or what have you—I might try to summarize by saying, "*Okay, here's my sense of what Professor Steinhardt is saying. He's saying, 'We had a problem because the universe has these particular characteristics that we couldn't explain and seemed unlikely to have occurred just by chance.' Then it turned out that what seemed to be a promising approach towards a solution to this problem, involved three separate issues. The first issue was that we had to make a very particular choice in order to launch this mechanism to work the way that it should. The second is that, it turns out that, even if one can somehow find a way to do that, it is far, far less likely to happen than we might imagine or hope. And the third is that, even if we could get it to begin and work in the way that we would like, it will never actually stop.*"

PS: Yes. And it will produce a multitude of patches with every possible outcome, which means that you can't assign a prediction to such a theory, because literally *every* physical outcome that is possible occurs an infinite number of times. So it's a theory that literally says that *anything* is possible.

HB: Right; and presumably not just that anything is possible, but that anything *will* happen.

PS: That's right, yes.

HB: So, again, if I'm this guy who's watching this I'd probably think something like, *Well, it's time to pack up and say that this approach clearly doesn't work—that doesn't make sense. That sounds like a big failure—not just a little failure—that sounds like a completely wrong idea.*

Moreover, I might actually start losing faith in the scientific process when I hear this sort of stuff. I might even think, *What are we paying these guys money for if they're coming up with answers like this? What are my tax dollars going towards if this is what these greatest minds of today are actually saying?*

So, let me ask two specific questions. The first is, *What would the response by an active proponent of inflation be to that?* The second is, *Is there starting to be a growing sense of frustration amongst the theoretical cosmology community in resonance with some of your concerns, or would you consider yourself more in the minority?*

PS: Okay, that's a good set of questions. You get a wide range of reactions. I think this is an interesting, sociological, scientific situation where an idea has become largely accepted by a combination of communities—astronomy, particle physics and cosmology—and the realization that it has the problems I've discussed has only dawned on them after this acceptance, so they've become committed to it.

The way science often works when it comes to a situation like this is that someone working in the field actually may not have thought deeply about, or even know about, all aspects of the idea. If I'm an astronomer who's trying to use the distribution of galaxies to measure how close we are to a flat universe alleged to be predicted by inflation, my measurement has a validity whether or not the theory is true. I might use the theory as an inspiration for why I'm doing the measurement, but I might not really examine it in detail.

There's a large number of people in the community who joined the field beginning in the 1980s. And even though in the

1980s, all three problems—the fine-tuning problem, the initial conditions problem and the multiverse—began to be understood and appreciated at that time, many people entered the field saying, *"Well, I'm going to assume that those problems will be solved, and I'm now going to work on this one particular issue."*

Now it's the case that several generations of students have gone by who have been educated that this theory works, but they're not taught about the initial conditions or multiverse problems. So to some degree, you have to remind people or tell them that there actually are these problems.

In the early years, I think many people, including myself—I was naive about it too—thought that these problems were going to be worked out, since we managed to get past the other ones.

In the early years, that was a reasonable point of view, but I think what has happened over the course of the past 30 years is that every attempt to solve those problems has failed, and it's failed in a way that makes it clear that the problem is much worse than we had thought.

For example, let's take the multiverse problem. Many people would admit, *OK, it's true that it produces an infinite amount of everything*, so most don't argue that the theory produces a multiverse. But then they'll say, *"Well, I'll now add a new rule to my story, which will tell me which patches of the universe count more than others"*. In other words, I'll come up with an answer for the question, *How do I weigh my different patches of the universe that come out of the multiverse to decide which one is more probable?*

It would be like if I gave you a bag of coins of dimes and nickels, and as long as it's a finite bag you could calculate and say definitively whether there's more of one than the other, and you could do various spot tests by taking samplings to say which is more likely than the other. But first you'd have to have a rule as to how you would do it.

So they would add a rule—a so-called "measure rule" or "measure principle"—that they hope will resolve this problem. Now, actually, that's a really big thing to be adding. By calling it "a measure principle", it sounds somewhat innocent, like they're adding a little thing, but the fact of the matter is that the fact that they're adding it is a sign that the theory has failed.

Inflation was **supposed** to give you what you wanted already, and it **didn't**. The problem was that, coming out of the Big Bang, you had no idea what to expect, and what we actually observed seemed unlikely. And now inflation has produced an **infinitude** of possibilities, which means that you're pretty well back to where you started. So now you're going to add a measure principle to solve this?

Imagine if you **did** find the right measure principle. Now it's not *inflation* that's done anything for you, it's the *measure principle* that's done all the work, because without the measure principle, we just agreed that there's no solution. So it's a *really big thing* to add such a measure, and it totally changes the theory.

Moreover, as you'd probably guess, different measures might give you different results. And that's exactly what people found: that different measures give you different things. But as it happens, much to their horror, it's *also* true that they haven't yet, even today, found a measure in which our *actual observed situation* ended up being the most likely.

So, for example, the most obvious thing to ask is, Are there more bubbles of our types or other types? And the answer is "other types," by an exponential amount, like 10 to the 10^{100}.

Then you might say, "*Well, let's focus on volume instead, because inflation is all about making volume. Is there more volume like us or not like us?*" Answer: **not** like us, by a factor of 10 to the 10^{100}.

HB: You can't even really use the anthropic principle either, it seems. That's one of the few times you can't even use that to save you.

PS: Well, that's often invoked with a combination of measures. The measures get more complicated, and every time they fail, somebody invents something more complicated, and sometimes those involve the anthropic principle.

But that doesn't help you either, because among the infinitude of multiverses are parts of the multiverse which are exactly like us, have the exact properties as us, but are younger. And they are exponentially more probable to occur than we are.

So imagine bubbles, or patches, which are just a second younger than us—it seems pretty reasonable that we could have existed just a second ago. But what if I told you that there are 10^{50} times more patches like *that* than like *us*? Then it's really hard to explain why we're as old as we are.

And bear in mind that I've separated out all the bubbles that *aren't* like us—I didn't even count those. So even the ones that are like us but just a little bit younger are more probable. And the reason is because, as time goes on and you're inflating more and more, you're producing more and more volume and that leaves room to produce more patches. So the longer you wait to produce your patch, the more room there is to produce such patches, and the more of them you produce.

What that means is that younger patches—that is, ones produced more recently—are always more likely than older patches. And we're an old patch: we're 13.7 billion years old. So one that's 13.6 billion years old turns out to be hugely more probable than we are.

HB: Okay, so I'm going to bring back my sceptical gardener again who's listening to all of this. I'm guessing that my sceptical gardener would say, "*Okay, it seems to me that there are two possibilities going on here. It might be the case that, if I talk to another scientist who's working in this field, she'll tell me that Steinhardt is all wrong—'**He's crazy, he's got his figures wrong, his calculations are wrong, it's all wrong; he used to be a good scientist, but now he's done, he's finished, he's over the hill**'—and proceed to show me a bunch of calculations to explicitly demonstrate why, exactly, you're wrong.*

*"Or she might say something like, '**Well, okay, he's right in what he's saying, but we'll get around all of that eventually**.'"*

PS: I think most of them would be in that second category. Most of the people who would describe themselves as cosmologists actually don't work on those problems, but I think they would say that the idea of inflation seems so sweet—much like the attitude I once had—that it's *got* to be right, so those problems will eventually be solved.

But they haven't tried to attack them, and they haven't followed the attacks—it's only a fairly small community that has struggled with this problem, a fairly small number of people out of all the people who work on cosmology.

So when they say that they think it can be solved, my sense is that they haven't thought very hard about it. The more you struggle with these problems, the more you realize that they're actually much worse than you think. Both problems, both the initial conditions and the multiverse, are actually much worse than you think because—well, we'd have to go through some examples.

HB: Well, they're bad enough already, it seems.

PS: Yes, that's right. And I should say that the reason why they think the idea is sweet is because they were taught it in a certain historical order, just as we discussed, which is, *Ignore quantum mechanics and you see this beautiful idea*. But once you add the quantum mechanics properly then the thing takes off in some direction you didn't expect.

And they like that first story so much, it's such a simple story—*stretch things fast and they become smooth*—they can't get over the fact that quantum fluctuations can take over the universe and produce this multiverse out of it.

In certain communities—I find this especially so when I'm in Europe—I've had this discussion where I'll say, "*Well, what do you think about the multiverse problem?*" and they reply, "*I **don't** think about it.*"

So I'll say, *"Well, how can you **not** think about it? You're doing all these calculations and you're saying there's some prediction of an inflationary model, but your model produces a multiverse, and so it **doesn't**, in fact, produce the prediction you said: it actually produces that one, together with an **infinite** number of other possibilities, and you can't tell me which one's more probable."*

And they'll just reply, *"Well, I don't like to think about the multiverse. I don't believe it's true."*

So I'll say, *"Well, what do you **mean**, exactly? **Which** part of it don't you believe is true? Because the inputs, the calculations you're using—those of general relativity, quantum mechanics and quantum field theory—are the very same things you're using to get the part of the story you wanted, so you're going to have to explain to me how, suddenly, other implications of that very same physics can be excluded.*

"Are you changing general relativity? No. Are you changing quantum mechanics? No. Are you changing quantum field theory? No. So why do you have a right to say that you'd just exclude it?"

HB: And again their answer to that is just, *"We just don't think about it"*?

PS: Yes. There's a real sense of denial. It's just, *"Somehow I just think there must be something wrong with that whole multiverse idea."*

I actually think that, indeed, there **might** be something wrong with the way we often describe the multiverse as patches of stuff. That may not be right, but the answer might be much **worse** than that when you do the full quantum mechanical calculation.

Before we continue, I should mention a particular example of the sort of attitude I'm talking about here. In March of 2014, there was this announcement by the BICEP2 experimental team at the South Pole that was making detailed observations of the cosmic microwave background that they had discovered polarization signals, which they attributed to being due to primordial gravitational waves consistent with what you'd get from inflation.

Now I just explained that the multiverse produces an *infinitude* of possibilities. But in spite of that fact, people declared that announcement as "proof of inflation." In fact, somebody even declared it as "proof of the multiverse," which is a pretty odd phrasing. I mean, we're talking about the leaders in the field here, saying peculiar sentences like, *"In the multiverse I wouldn't have really expected to see those gravitational waves, so it's great that we were lucky enough to be in the patch that has them."*

HB: I would have thought that you could just declare anything that you see as evidence of the multiverse.

PS: Well, yes—so actually, a more extreme version of that question is, *Is there anything you could observe that would tell you that inflation is wrong?*

And, again, for many of the leading proponents of the field, the way they answer that question is to say, "**No**. *Inflation is so flexible that no test or combination of tests can possibly disprove it.*"

In fact, in their view it has three degrees of flexibility: there are those initial conditions that I'm allowed to fiddle with, those parameters that I was allowed to fiddle with, and then there's a multiverse that I'm allowed to fiddle with. So according to them it's super-flexible.

And so my response, *"Okay, then, doesn't that mean you concede?"*

And their reply is, "**No!** *What's wrong with that?*"

So we kind of have a philosophical difference here. I would have said that a theory that is not empirically testable is not scientific—you can't eliminate it—to which they'd say, "No, *through the tests, we'll just figure out which of the inflationary scenarios and parts of the multiverse we happen to be in.*"

HB: That sounds like such extraordinary question-begging that, one possible solution, on a sociological level, would be to force people to take philosophy courses at an earlier age. But that's a whole separate issue.

238 CONVERSATIONS ABOUT ASTROPHYSICS & COSMOLOGY

PS: Well, I've actually suggested that.

HB: You have?

PS: Yes, I've been saying that one thing I think we would benefit from would be sensible, basic philosophy, which most physicists don't take. Unfortunately, I haven't been able to encourage any philosophers to come and weigh in on this—they have their own issues that they're interested in, I suppose—but I think we would actually benefit from some basic philosophical education.

HB: Interesting. But I cut you off with the BICEP2 story. I understand that there's more going on here than just enormous question-begging.

PS: Yes. What happened was that people were ready to declare inflation as having been *proven* on the basis of this observational result that had been declared. And *then* we discovered that what had been observed was *not*, in fact, primordial gravitational waves, but a polarization signal due to dust within our own galaxy— effectively, pollution of the signal.

So you would think, *If you just declared* **victory** *on the basis of the discovery of them, doesn't that mean that you have to declare* **defeat** *on the fact that you* **didn't** *see them?*

And the immediate response of the proponents of inflation was, "*Absolutely not. Our theory is flexible enough so that we can show that too!*"—and we immediately got a litany of papers saying, *Here's how we'll do that.*

I think this very public exposure has at least gotten some fraction of the community to realize that there really are these problems of ultra-flexibility and non-predictability to the theory.

HB: How large a fraction are we talking about here, roughly speaking?

PS: I've been to a number of meetings in which proponents have tried to sell hard the idea that inflation is still in good shape, and occasionally there are these things we do in conventions where people take votes like, *Who believes that inflation has been proven?* or *Who believes that inflation is the only idea?*

I think that if you were to have asked that question in March of 2014, almost everybody would have raised their hands saying that they believed in inflation—but now, at least based on my latest, anecdotal experiences, people actually don't raise their hands, except for a few proponents. I think that means that many people are beginning to rethink the issue. I don't know how long that will maintain itself, but I think it's at least a shift in the right direction.

As I said, part of it is that the story you read in text sounds so simple that it's hard to believe that there could be any problems with it, but it takes more work to see these problems. But once you see them, they absolutely take over.

HB: I'm clearly no expert, but my own vague sense—unaware of the complications of the multiverse that you've been describing— was something like, *Well, if they actually do detect this primordial, gravitational radiation they're talking about and they can observationally make sure that it's what they say it is, then that seems to be some very strong evidence of that model. Otherwise, you'd have to say that there must be some other model that predicts this particular thing.*

PS: Which, by the way, is possible. In other words, there's nothing about these gravitational waves, even if they *were* to be observed, that would explicitly say "inflation" on them.

HB: That's what I was getting to. So my thinking was that they were premature in announcing victory because they thought that this signal meant one thing when it actually means something else, because they didn't properly calibrate their equipment and the observation was due to dust, or whatever. So my naive sense was,

OK, they were premature. They jumped the gun and held a press conference and all that, and that's what all the fuss was about.

But that's very different from this idea that it *doesn't actually matter* what is actually detected, that in fact they can justify their theory *whatever happens.* It seems to me, from what you're saying, that clear evidence of detection of these gravitational waves can be used to justify inflation just as much as a non-detection of them.

PS: Oh yes, they've already decided that: they've already written the papers that say things like that. They've already written papers that celebrate the fact that, *In my favourite model, I can adjust a parameter and get any answer that might be observed by experiment.*

To which my response is, *That's terrible, that's not a good situation, that's a very unhealthy situation.*

This is reflective of a certain fraction of the community that is very heavily emotionally invested in the inflationary idea, but as I said earlier, I think that there's also a large and growing fraction, which has seen this activity and has the same reaction that you do: *This isn't what we signed up for.*

I think they would like to see something else develop—some other ideas—but there isn't something on the table now that is fully developed and fully thought-through that they find convincing.

HB: Right, but that's an opportunity to do work, right?

PS: It's a *tremendous* opportunity. I try to tell students that this is exactly where you want to be coming through, because clearly this idea is wrong and has to be replaced. Personally I have some ideas that might replace them, but they're probably not right either, so we need some new ideas.

So, yes, it's a wonderful opportunity, it's a way of really getting past all these years of going in the wrong direction. When you see a crowd going off in a direction and you know it's the wrong one, you should definitely be headed the other way.

But I'm sorry to say that not many people do—not yet anyway —but I do think that the intellectual and scientific force is there to make real progress.

As I've been saying: inflation has some very hard problems with it, and nobody is claiming to have solved them—certainly no one has convinced others that he's solved them.

HB: Those strike me as pretty hard problems to solve. I mean, from what you were saying, I'm not even sure it's logically possible to solve them.

PS: I would actually support that. As one grapples with it, you realize that they're completely sticky.

Questions for Discussion:

1. To what extent is it justified to continue working on a theory while still recognizing that many fundamental issues need to be resolved? Where should one draw the line and recognize that it's time for a completely different approach?

2. Can one make an objective distinction between proponents of inflationary cosmology clinging to their theoretical framework independent of any experimental results and adherents to a revealed religion?

3. Should all physicists have to take history and philosophy of science courses as part of their education?

4. Did the media do a responsible job in conveying the full picture of the controversy associated with the BICEP2 announcement and later retraction?

VI. Bouncing Back?

Considering a key change

PS: So how do you go forwards? You need to say to yourself, *"Okay, I **had** an idea, I **thought** it worked, but it **failed** in particular ways. So obviously I have to go back and change something that I had assumed"*.

So let's go back. What did I assume to make the theory in the first place? Well, not very much.

We assumed there was a Bang, we assumed there was general relativity or something like general relativity that described an expansion of the universe that was initially hot and began in some sort of random wild state, and then we assumed that—somehow—inflation took over and got things started. But that whole line of thinking has now failed, as we've discussed, because inflation needs its own special start and because it produces the multiverse.

So what do we need to do next? Well, *what* can I change?

I *have* to have gravity, so how about the idea of a Bang being the beginning? What if the Bang, instead, were a sort of Bounce? That's to say, it isn't the beginning, it's just a transition from some kind of, let's say, contracting phase to some kind of expanding phase.

Could that possibly gain you something? Well, as you think about it a little bit, it automatically begins to do some really good things for you. One of the problems we had in explaining why the universe was so uniform was that, according to the Big Bang theory before we developed inflation, distant regions that we see in the universe hadn't been in causal contact before.

And the idea of inflation was to say, *Actually, they were much closer together in the past and then they were thrown apart during inflation, so these different regions actually were in causal contact before.* But that, unfortunately, led to this failure mode: that whole line of reasoning about inflation.

But now if you assume that the universe didn't begin at the Big Bang, you've got lots of time beforehand for different regions to come into contact. So you immediately get around the contact problem.

Another thing you immediately get around is the flatness problem. How is inflating solving the flatness problem? It was solving it, not by setting it to zero, but by suppressing it by a huge amount, by this super-stretching. Then, after inflation ends, it begins to grow again, but you've suppressed it so much that, today, it's immeasurably small. That was the concept there.

Well, it turns out that if the universe is contracting, that automatically flattens the universe. You can see that in various ways, but since we're not going to write equations on the board, let's instead just think about what the original flatness problem was.

The original flatness problem was, if you began from a Big Bang that was slowly expanding—not inflating—the universe should become more and more curved; and so, today, you should see a very curved universe instead of the pretty flat universe we find ourselves in. That was the flatness problem.

Now roll that film backwards—the universe, which is very curved, is slowly contracting and it becomes flatter and flatter, so the same physics that we imagine should do the curving on the way out does the flattening by contraction on the way in.

In other words, as long as I have enough contraction, then after the bang, there's only been a certain amount of time since the "bounce," as we're now thinking of it, and it just hasn't expanded enough yet for the curvature to be visible. That would solve that problem.

So you can see that, just by changing that idea, you immediately solve two of the problems that inflation was designed to solve. You also open up the time domain, which enables you to address the initial conditions problem. Now you can imagine a very large universe that is very classical, in which quantum physics isn't very important where you could have set those initial conditions.

I haven't gotten into any specifics here, but as you begin to think about this idea as a theorist, you realize that you immediately gain something just by replacing a bang with a bounce.

Of course you have to do a lot more. You want to explain a lot of details: why is the universe today not just flat, but also so smooth? Why is it so isotropic? How do you get those density fluctuations?

You can fill in those blanks—probably in more than one way—with a sequence of events that would have occurred during that period of contraction.

HB: As you rightly say, Paul, this isn't the best forum to go into those details, but it's a natural point for me to mention that you've written a book, *Endless Universe*, detailing your ideas on this topic for a general audience.

Thanks very much for taking the time to explain your ideas and concerns so clearly, Paul.

PS: My pleasure.

Questions for Discussion:

1. Might there be negative aspects of transforming from a "bang" to a "bounce"?

2. Is it possible that there will be some cosmological mysteries that we will never be able to solve? If so, why do you think that is?

3. Do you think that some areas of physics are more open-minded and philosophically-sophisticated than others? If so, what might account for the difference in attitudes across sub-disciplines?

*4. How common is the notion of a "bounce" rather than a "bang" throughout the scientific community? Readers interested in this topic are referred to the Ideas Roadshow conversation **The Cyclic Universe** with Roger Penrose, where he describes his theory of Conformal Cyclic Cosmology which is also predicated upon fundamentally rethinking our conceptions of a Big Bang.*

Continuing the Conversation

Readers are encouraged to read Paul's book, *Endless Universe*, which goes into considerable additional detail about some of the issues discussed in this conversation.

Those who enjoyed this discussion are also strongly recommended to read an entirely different Ideas Roadshow conversation with Paul, *Indiana Steinhardt & the Quest for Quasicrystals*, where he describes his fascinating story of discovering the world's first natural quasicrystal.

The Cyclic Universe

A conversation with Roger Penrose

Introduction

A Hidden Audience

Does it make sense to use the vehicle of a popular science book to put forward one's own detailed theory of modern cosmology?

Obviously not.

After all, cosmology is a highly technical, deeply abstract field of study that represents the culmination of much of modern physics. In order to fully comprehend its subtleties, one must have a thorough mastery of general relativity, differential geometry, thermodynamics, quantum field theory, and a good deal else besides.

Frankly put, it is virtually inconceivable that a non-specialist could somehow navigate his way through the rigorous arguments to have a genuinely clear idea of what is actually being proposed. In order to reach a broad audience, many highly technical issues will have to be glossed over, which will clearly trivialize the ideas beyond recognition, eviscerating them of any scientific content whatsoever.

And then there's the fact that a well-defined, international community of scientific experts already exists for precisely the purpose of evaluating new ideas: there is a welter of well-respected journals in which to publish original work, and no shortage of scientific conferences to attend to speak to one's professional colleagues. Any attempt to somehow circumvent this process and publish one's pet theory in a popular book is hardly destined to be enthusiastically endorsed by the scientific establishment.

For most of us, then, the idea of producing a popular book about our iconoclastic cosmological views is little short of a terribly bad idea. The general public won't be the slightest bit interested, and the scientific community will have you for lunch.

But then, most of us aren't Roger Penrose.

Penrose's scientific credentials are, of course, unimpeachable. One of the world's most accomplished mathematical physicists and a specialist in general relativity and cosmology, his litany of intellectual achievements is nothing less than outstanding. He is the originator of the singularity theorem in general relativity— later extended to cosmological scales in his collaboration with Stephen Hawking—twistor theory, the cosmic censorship hypothesis, Penrose-Terrell rotations, Penrose tilings, spin networks, and much more. His work on black holes that began with his singularity theorem was recognized in him being a co-recipient of the 2020 Nobel Prize. He is, quite simply, beyond whatever slings and arrows the scientific establishment might care to hurl at him.

He has also, surprisingly, become a fantastically successful popular writer. Why surprisingly? Because he has long eschewed the common dictum that the public can't handle mathematical details. His 2005 opus *The Road to Reality* was firmly ensconced on the bestseller list for months on end, despite both its daunting size and equally daunting number of equations.

So Roger Penrose, clearly, is different. But what has he got to say this time? His popular book, *Cycles of Time*, explains his theory of Conformal Cyclic Cosmology—CCC—to a broad, general audience. Well, what's *that* all about?

In essence, it proposes that cosmology should properly be regarded as an infinitely repeating series of "aeons," with the far distant future of one aeon mapping on to the Big Bang of another one, thereby producing endlessly repeating cycles.

Given the seemingly endless proliferation of bizarre-sounding theories of fundamental physics out there, you might well think that this is just another in a long line of provocatively speculative ideas cooked up by a restless physicist anxious to distinguish himself by the sheer power of his creative, science-fiction-like energies.

But you'd be dead wrong.

Because what's essential to understand is that Roger's latest ideas hardly represent a departure from a fundamental cosmological paradox that has long haunted him: Why is the beginning of the universe in such a remarkably smooth state?

> *"It's often claimed that there are these big mysteries about the universe—**What is dark matter? What is dark energy? Where do they come from? What are they doing?**—together with the values of all sorts of other parameters that seem completely mysterious. But they **never** mention the Second Law of Thermodynamics. For some reason, when people enunciate the various problems of cosmology they don't even ask, **Why was the Big Bang not only such a state of low entropy, but a state of such low entropy in a very strange way—that it singles out gravity as the one thing that is not taking part in this thermal state?***

> *"Gravitation is aloof from everything else going on, and only gradually does it get brought in and produce the concentrations of stars and thermonuclear reactions and all sorts of things. But the key reason why the entropy was low was due to gravity, because it's taking advantage of this reservoir of low entropy since the gravitational field was not taking part."*

For those who claim that the theory of cosmic inflation solves this deep puzzle, Roger is swift to disabuse them. He admits to not liking inflationary cosmology on "aesthetic" grounds, feeling that it is too ad hoc. But that's not his main issue with the theory.

> *"My principal concern is that it didn't actually **explain** what it was supposed to explain. And the reason for that is that, in my*

*view, the **bigger** problem is the Second Law of Thermodynamics. The spatial uniformity that inflation is designed to address is part of this problem: How is it that the gravitational degrees of freedom were not activated? Putting inflation in doesn't solve it.*

"In fact, I can show you why it doesn't solve it. It's so remarkably simple I can't see why others haven't been copying this idea endlessly. I'm going to turn the universe upside down. Why? Well, because I'm now going to think of time as going in reverse back towards the Big Bang.

"Entropy will increase through gravitational clumping. There will be irregularities as the universe collapses, and these will increase and form black holes: a huge, horrendous mess.

"That is, almost certainly—with fantastically likely probabilities— what our universe would do. And if we had been in this unbelievable messy situation to begin with, then inflation wouldn't do anything for us at all. The 'unbelievable mess' would have been a state of enormously high entropy (in terms of the gravitational degrees of freedom), and inflation, being a time-reversible dynamical process acting in accordance with the Second Law of Thermodynamics, wouldn't be any use at all: it would just spread out the clumps. So it's really no explanation to the question of why our universe is so uniform."

Roger has long been preoccupied with this issue, and believes that a key piece of the puzzle can be properly addressed by closely examining aspects of a particular type of geometry—so called conformal geometry.-

In fact, a careful examination of Roger's research orientation, from Penrose diagrams to twistor theory to the Weyl curvature hypothesis, demonstrates a consistent belief in the importance of conformal geometry.

CCC then, is hardly some new, fantastic theory that comes straight out of left field, but rather a logical culmination of many of Roger's remarkably fruitful scientific beliefs, intriguingly extending these to incorporate fascinating ideas about dark energy, dark matter and information loss in black holes.

Unfortunately, however, many contemporary physicists summarily dismiss these ideas, looking askance at CCC on experimental grounds. Because one of the main predictions of the theory is a specific ring-shaped pattern in today's cosmic microwave background (CMB) caused by giant black-hole collisions from the previous aeon.

Some believe that such ring-shaped patterns are nowhere to be found, while others admit that they *are* there, but are not at all particular to CCC. Of course, time will tell.

But as important as experimental confirmation is to any scientific theory, even more important is a recognition of the need to address fundamental, underlying issues. Because even if CCC is only somehow half-right, or not right at all, that hardly means that Roger's long-standing cosmological concerns are misplaced.

> *"You have this amazing piece of evidence of a thermal state, which means a state with **maximum** entropy. To me, that doesn't make any sense. Why aren't people saying, '**WHAT? You go back to the state of smallest entropy and get this thermal equilibrium state, which is a sign of maximum entropy?**' Why aren't they scratching their heads and saying, '**This is madness!**'*
>
> *"I don't understand why they don't worry."*

Which brings us, in a roundabout sort of way, to the subject of why Roger wrote *Cycles of Time* to begin with.

> *"When I wrote **The Emperor's New Mind**, I went into this business of the low entropy state of the Big Bang and the fact that gravitational degrees of freedom had to be suppressed. I'd been lecturing about this for several years before that, and nobody paid the slightest bit of attention to it. But after I'd done it in **The Emperor's New Mind**, a lot of my scientific colleagues wrote to me and said, '**Oh, that's interesting. Now I see what***

you're saying'. It did actually get across to people in a way which writing articles in journals and so on didn't."

Might **that** just explain the Penrose publishing phenomenon? Could it be that, through the guise of writing for the layperson, Roger is really addressing the global community of theoretical physicists? And might it also be that, despite their official professional disdain for popular science books, a significant fraction are going out and buying his?

The Conversation

I. Inadvertent Success

An inside look at a mysterious publishing phenomenon

HB: I'd like to talk a little bit about the way you communicate your ideas to the general public. You've written many books for the general public, and these books have been extremely successful. Quite frankly, I'm not sure what to make of that.

On the one hand, I think it's marvellous that people are motivated to read your books and be engaged in the theatre of ideas.

But the way you write these books is very unique: they are conceptually demanding and technically demanding. They require a tremendous effort by the reader, both intellectually and in terms of time. In many ways I think this is very positive: a sign that many people outside the professional community are excited by theoretical physics. But I can't help but wonder how many people are buying these books and *not* actually reading them, sticking the Penrose authority on the bookshelf as a sort of "physics bible".

It seems to fly in the face of so many contemporary trends. There's this famous anecdote when Stephen Hawking was told by his publisher that every mathematical equation he would insert into *A Brief History of Time* would halve the number of sales, so he didn't put any in at all.

That certainly doesn't seem to hold true for **your** books. Or perhaps had you left all equations out they would have outsold any book in human history.

RP: I know. I worked it out once. I took a sample of 20 pages, counted the equal signs and from that worked out roughly how many equations there were in *The Road to Reality*. And I worked out that if there had been a book sold for every proton in the observable universe, that would not **nearly** have been what my sales would have been, had I not put equations in.

HB: Just the very fact that you worked that out...

RP: Well, it was for lectures I was giving to promote the book.

But I think that ascribing any kind of motive is probably a mistake. It's true that there's a sense that I'm aiming at a readership that is not addressed: on the one hand you've got technical books and papers, while on the other hand there are popular books which don't really go into the equations, limiting themselves to analogies and so on. So there's a gap. And there's not much in that gap.

So my claim, when talking to my publishers and editors to get them to take these things seriously, is that there is a gap. How big it is, I don't know. But the books seem to sell reasonably well. So I think there's something there.

I really have no idea what people get from these things. That's not really the reason why I do it: I do it because I want to do it somehow.

HB: *The Road to Reality* is a comprehensive, almost encyclopedic, summary of physics, which must have taken a tremendous amount of work. I've long wanted to know what motivated you to write it.

RP: Well, you see, it's not correct to talk about what motivated me to write that book, because I *wasn't* actually motivated to write it. Events just creep up on me. It's not that I do it deliberately: I just can't stop it somehow.

I had written *The Emperor's New Mind*, and *Shadows of the Mind*, and I won't go into the detailed history of those books now, but essentially I had wanted to get a certain idea across about the mind combined with an excuse to write something vaguely popular about physics.

And a number of people had said to me, including my editor, that it would be awfully nice to take out all the contentious stuff about the mind, and leave just the descriptions about physics. And I thought, *"Yes, that's not a bad idea; it sounds pretty easy."*

So I mentally took out my scissors and cut out all those pages. But then the thing just fell to pieces: it didn't have a thread to hold it together. So what was going to be the thread that holds it together? The search for the laws that govern the universe.

But then I thought, *I'll have to bring in various things which aren't normally talked about; there's this and this and that....and then if I'm going to talk about this later on, I'd better introduce it earlier in some form...*

Soon enough I thought, *God, that's looking like an awful lot*, but I kept putting things together, and stupidly I didn't stop and tell my editor, *"Look, it doesn't work."* That's what I should have done. But instead I thought, *There's quite a bit of stuff here. It'll be about 500 pages, quite a big book.*

But it wasn't 500 pages, of course. It grew and grew and grew, and eventually it doubled in size. Had I known it was going to be 1,100 pages, or whatever it was, I wouldn't have written it.

In short I hadn't intended to write *The Road to Reality*. That's the answer to your question.

HB: But having written it, you must be quite pleased by the response. It seems to have sold an enormous amount of copies.

RP: Well, again, it's serendipity or something. When I wrote *The Emperor's New Mind*, I originally had in mind that I wanted to get this mind business across because I hadn't seen that particular point of view expressed. But I had also hoped, in describing all the physics, that there would be young people who might be inspired to do physics. So I sat back after having written it, wondering if anyone would write to me about it.

And indeed I did get quite a few letters, but they were mainly from old, retired people. A few young people did respond—one in particular who became a very well-known singer—and I was very pleased with that. So perhaps it did have a little bit of an effect on people who did physics.

The other thing that I hadn't anticipated, apart from the retirees, was that it would interest people in other areas of science—in particular Stuart Hameroff, an anaesthesiologist in Arizona. He's not just interested in putting people to sleep before operations, but he's also interested in what on earth he's doing when he puts people to sleep.

It turns out that he had his own ideas about these microtubules, which I'd never heard of. One of the ambitions of writing *The Emperor's New Mind* was that, although I didn't know what it was that could support these quantum coherent things in the brain, I thought that perhaps by the time I'd written the book I would have found out. But I didn't.

Suddenly, though, Stuart wrote to me out of the blue, and I learned about these things. He had actual pictures of them, so it was clear that they must be real. And it seemed to me there was a much better chance of preserving some sort of quantum coherence through this than anything else I had seen before. That led to a collaboration that is still with us.

Which is to say that it's not just young or old people, but more generally people working in different areas of science. When I wrote *The Emperor's New Mind*, I spent some time on the puzzling issue of the particularly low entropy state of the Big Bang and that the gravitational degrees of freedom had to be suppressed. I had been lecturing about this for several years, but nobody seemed to be paying the slightest attention to it.

But after the book was published, a lot of my scientific colleagues wrote to me and said, *"That's interesting;* **now** *I see what you're saying."* Somehow it did actually get across to people in a way that writing journal articles didn't.

HB: That must have been very encouraging. And, to some extent anyway, that probably increased your motivation to write further popular works.

RP: I think so, yes. There are sometimes advantages to writing something in science that's a little broader. In physics people tend to get focused on one area and they may not have seen what the problems are in the other areas. The foundations of quantum mechanics is one example of this: many people just leave things to the philosophers to worry about the correct interpretations, when I'm trying to say that there's a real problem there.

Questions for Discussion:

1. Which of Roger's popular books did you think were the most intriguing and why?

2. Is a popular science book an appropriate place to introduce new scientific theories and interpretations?

3. In what ways has an increased level of scientific specialization resulted in a situation where it is much more difficult to "see the forest for the trees"? Might there be any specific ways to significantly push back against these forces of specialization?

4. To what extent are the motivations of publishers of popular science books aligned with those of the authors? To what extent are they aligned with the interests of the general public? Are there too many of certain types of popular science books? Are there particular types of popular science books that you would like to see more of but don't yet exist?

II. Considering Entropy

Minimum, maximum and both at the same time

HB: So today we're going to talk about cosmology, and your particular views on the subject. And the story begins with entropy.

RP: Yes, well, it's a thing that has puzzled me for quite a long time. I should explain what entropy is. Entropy, in rough terms, is a measure of randomness.

There's this thing called the Second Law of Thermodynamics, which can be phrased in the form of saying that entropy is always increasing with time, or at least never decreases with time: it's getting bigger and bigger. Which means that the randomness or the disorder in the universe is increasing with time.

And you might say, "*Well, that's perfectly natural: you spill this coffee on the floor, then it gets more disordered, the entropy's gone up and so on. Yes, it's happening all the time.*"

You may worry a little bit because it seems that people, while they may behave in disorganized ways, are actually very organized structures. And how did that come about when you have a Second Law that says that things are getting more disordered?

But on the whole, as I said, it's natural to think that things are getting more and more random as time goes on. But if you think of this in the *backwards* direction of time—I'm not saying anything different than the Second Law, I'm just saying it a different way—then entropy will naturally *decrease* as you go back in time— things get more and more ordered. And if you go back and back and back, where do you end up?

Well, the current picture of the universe is that it started with this thing called the Big Bang. The Big Bang, therefore, must be a state of extremely low entropy, because the entropy is going down and down as you get closer to the Big Bang back in time.

Yet, what is the Big Bang? Well, it's this big walloping explosion. And what's the most impressive evidence that the Big Bang ever took place? It's this thing called the cosmic microwave background (CMB) that gave rise to the Nobel Prize on two quite separate occasions. It's a tremendous achievement to have observed it and analyzed it in the detailed way that people have been doing.

And what we see—well, we don't actually see it with our eyes, because it's electromagnetic waves of very low frequency—microwaves, just like we have in our microwave ovens—is radiation coming very uniformly from all directions across the whole sky that has a very particular character—so-called "black-body radiation"—which was explained by Max Planck in the beginning of the 20th century in a very important piece of work which actually started quantum mechanics.

HB: Let's back up for a moment. I can imagine that some people might ask, *Why are you physicists so convinced that there was a Big Bang anyway? Maybe there wasn't a Big Bang.*" You mentioned the CMB just now, but perhaps you can spend a moment to explain how that links up with our conviction that there is such a thing as a Big Bang.

RP: Well, that there should be such radiation is consistent with the picture that has been promoted for a Big Bang. That is, there was this extremely hot state in the beginning, which is called a "thermal state".

In fact, there's a little bit of a problem here because thermal equilibrium means that things are in a final state that doesn't change.

You might imagine a box with stuff in it, and after a long period of time it settles down into this thermal equilibrium state, a state of maximum entropy.

Now, of course the Big Bang wasn't like that because it was a great explosion, but it was very much like that in all other respects: you could have a box that could expand in such a way that keeps the entropy constant—a so-called adiabatic expansion. That's the kind of picture that people have for the Big Bang.

And that state should be a thermal equilibrium state. That's what people were expecting. And, lo and behold, we find that this radiation has this thermal character to an enormous precision. It agrees with this Max Planck curve I was speaking about a moment ago to a fantastic degree, much better than anybody could produce in the lab. So you have this amazing piece of evidence of a thermal state, which means a state with **maximum** entropy.

HB: So how does that make sense?

RP: Well, to me, it doesn't make *any* sense. Why aren't people saying, "**What?** *You go back to the state of **smallest** entropy and get this thermal equilibrium state, which is a sign of **maximum** entropy*"?

Why aren't they scratching their heads and saying, "*This is madness!*" I don't understand why they don't worry.

I think there is this sort of the feeling of, "*Oh well, because the universe is expanding, it's a different sort of situation...*"

Well, it's got nothing to do with that. This was made very clear by Richard Tolman, a very great American mathematical physicist who understood these things very well. In short, it *was* a thermal state and it was *expected* to be. Now, **there's** a contradiction.

To understand what's really going on, one has to take into account what one is actually seeing in this radiation. What you're seeing is photons—particles of light—which have come from this great mess of matter and radiation in this sort of equilibrium (also expanding, but basically in equilibrium).

HB: Just a brief clarification: what we're actually seeing is not these photons *directly* from the Big Bang, we're seeing them from this "surface of last scattering", we're seeing them at a particular time which happened a few hundred thousand years after the Big Bang.

RP: That's right. I've simplified things somewhat. People tend to call this cosmic microwave background, or CMB, "the flash of the Big Bang cooled down by the expanse of the universe".

Well, it's not that, as you said quite correctly: it's what happened 300,000 years after the Big Bang. But you couldn't see into what happened before that time because the universe was opaque to us and the radiation didn't get out.

That's an important point, but of course it doesn't in any way explain what we're talking about here, because that, already, would have to be a very low entropy state compared to where we are now.

So how could that be? Well, the reason is that what we've been looking at so far is matter and radiation in thermal equilibrium, but that's not all the physics. That's most of physics, but it leaves out one important thing: *gravity*.

Let me use the following transparency to show you what I mean.

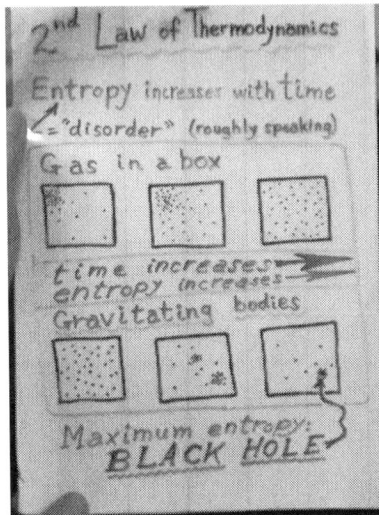

So here I'm illustrating the Second Law of Thermodynamics, with time increasing as we move to the right. Suppose you have a gas in a box, with the gas initially tucked up in the top-left corner of the box, as represented by the left-most picture of the first row of the transparency. Then you open up the doors of the compartment and the gas spreads out throughout the box, which represents an increase in entropy, so that our last picture of the gas spread out throughout the box is a high entropy state compared to the initial state where it was all concentrated in one region.

Now we move to the second row of pictures, where I want you to imagine a much larger scale situation. Now, instead of a lot of molecules in a box, we have a lot of stars in a great galactic-scale box. You then let time elapse—moving to the right along the diagram— and because of gravity they will start to clump and produce irregular distributions.

So entropy is still increasing as we move to the right along the row, but now it looks very different. Both scenarios are consistent with the Second Law, but the pictures you get are quite different.

In the first row of pictures with the gas in the box, we have a non-uniform state becoming a uniform state, whereas when gravity comes into the picture—in the second row of pictures—an initially uniform state becomes a non-uniform state.

And what we are seeing in the early universe is a combination of gravitational physics and the rest of physics, but in a uniform state—the last picture on the top row and the first on the bottom row.

And as I said, we find that not only is this radiation very much satisfying this Planck black-body curve I mentioned earlier—indicative of a state of thermal equilibrium—but it is very, very uniform over the whole sky.

HB: So perhaps one way of reconciling this entropy puzzle you mentioned earlier, of how the early universe could be regarded as being in both the highest and lowest possible state of entropy, is to regard gravity as being somehow "turned off" at this very early stage, so that we have a sort of thermal equilibrium— the top-right picture in your diagram—but the entropy is still very low because gravity somehow hasn't yet begun to do its thing.

RP: That's a fair way of looking at it.

I should say that historically, one of the reasons why people didn't worry so much about this is related to the way that relativistic cosmology developed. Einstein produced the equations of general relativity, and then people started to solve the equations—particularly Friedmann, Lemaître, and various others who solved the Einstein equations in a cosmological setting.

Now, in order to solve complicated equations like Einstein's, you need to make simplifying assumptions. And one of these simplifying assumptions is to assume that the universe is absolutely uniform spatially: it's more or less the same here as it is over there, the same in one direction as any another— homogeneous and isotropic.

You put those assumptions in, and then you can solve the equations. And that's not only the way cosmology started, but the way people are *still* doing it: that's the way people talk about cosmology most of the time.

But the trouble is that we've built in from the very beginning this initial uniformity of matter: we've *assumed* it. And people have forgotten that this is an absolutely **enormous** assumption.

It's worth stating, by the way, that it's clearly not the case that things are *always* homogeneous and isotropic. The reason we're here right now, for example, is that life arose because of the particular conditions that we have on this planet combined with the effect of the sun. People say, *"The sun's a wonderful thing: we get energy from the sun."* But that's not really right.

It's certainly not the main point, because the energy coming in from the sun in the daytime goes back again into space at night. So it's not so much that we simply get *energy* from the sun, because if that was all there was to it, the earth would just keep getting very much hotter and hotter as a result of this solar energy, which is not the case because this energy is reflected away into space (I'm not talking about global warming here, which is another thing altogether).

The key point is that it isn't the *energy* per se, it's the *entropy* of the radiation from the sun. We get it in a low entropy form—high-energy photons—and throw it away at night in a high entropy form —significantly more infrared, low-energy photons—that spreads all this randomness about.

And this low entropy radiation we get from the sun is attributable to the fact that the sun is a hot spot in an otherwise very dark sky. In other words, it's a case where things are very *inhomogeneous.*

If the sun's temperature were somehow spread out all over the sky, it would be completely useless to us. So it's not just the sun in itself that is important here, it's because we also have the cold background of the sky. This combination produces this cycle on earth: a relatively small number of high-energy photons comes to us from the sun and is eventually turned into a greater number of low-energy photons going away from us back into space, with all the entropy—all that randomness—transferred into those photons that the earth sends back into space.

The plants make use of this difference in entropy, and we make use of the plants and build up our structure. So that is how humans can be such incredible ordered systems, because we're using this low entropy from the sun.

HB: So this is an answer to the question you raised earlier: How is it possible that the occurrence of humans, such ordered and complex structures, can gibe with the Second Law of Thermodynamics? We're absorbing this low entropy from the sun.

RP: That's right, yes.

Questions for Discussion:

1. Why, exactly, does Roger say "If the sun's temperature were somehow spread out all over the sky, it would be completely useless to us"? Could you explain why this is the case to your neighbour? How is this related to the notions of thermal equilibrium and "black-body radiation" that Roger mentioned earlier?

2. How is the notion that "gravity is always attractive" related to our understanding of the Second Law of Thermodynamics?

3. Is it possible that when you open up the compartment containing a gas the molecules all randomly end up clumping together in one corner instead of dispersing in accordance with the picture at the upper right of Roger's diagram? If it's technically possible, how likely do you think it would be? To what extent does this concept redefine the notion of "possibility" in terms of "statistical likelihood"?

III. Mysteries and False Explanations

Dark issues and inflationary hyperbole

HB: OK, so let's return to cosmology now. There is this mystery concerning entropy and the initial state of the universe that we're obviously going to return to.

But more generally, we know that there is this cosmic microwave background radiation (CMB) that presents very strong evidence for a Big Bang. We know that this CMB radiation has this black-body spectrum and it's very isotropic and homogeneous throughout the sky, although people have very successfully examined the tiny deviations from that and been able to do some remarkable things, which I'm sure we'll also get to in a moment.

But there are some other things that we know too from observations: that there are these things called "dark matter" and "dark energy".

RP: Yes. Well, it's often claimed that there are these big mysteries about the universe: *What is dark matter? What is dark energy? Where do they come from? What are they doing?*—together with the values of all sorts of other parameters that seem completely mysterious. But they *never* mention the Second Law of Thermodynamics.

For some reason, when people enumerate the big issues in cosmology, they don't even ask, "*Why was the Big Bang not only such a state of low entropy, but a state of such low entropy in a very strange way—that it singles out gravity as the one thing that is not taking part in this thermal state?*"

Gravitation is not thermalized: it's aloof from all the other goings on, and only gradually does it get brought in and produce the concentrations of stars—and our sun, of course—and thermonuclear reactions and all sorts of things.

But the main reason why this initial state of the universe has such low entropy is because of gravity. That's the key reason why the entropy was so low at the very beginning: there is this reservoir of low entropy that results from the gravitational field not taking part.

HB: So, oddly, most people don't seem very concerned about this, but they *are* talking about dark matter and dark energy. Why are they talking about that? What's that all about?

RP: Well, let's start off with dark matter, because that was first discovered.

Our galaxy is held together by gravity. It's rotating, and you can work out how much matter should be present for the amount of rotation we observe. Now this applies not just to our galaxy, but all the galaxies around.

And people started to get very puzzled by the fact that there didn't seem to be enough mass in the stars we could see to account for holding the galaxies together for the observed amount of rotation.

HB: When did people first notice this?

RP: I believe it started with Zwicky in the 1930s. It is quite an old problem, but I suppose people really only started to worry when they could do the calculations accurately enough to realize how much matter was really missing.

It was called "the missing matter", but there were all sorts of names for it. All these names, to me, are bad names. "Dark matter" is also a bad name.

HB: Everything is "dark" now. "Dark" just seems to be the name for *We don't know what's going on.*

RP: Yes, but you see there *is* actual "dark matter," which is dust. But what people call "dark matter" is actually "invisible matter". But the story with the cosmological constant—dark energy— seems even worse to me. Perhaps I'm being a little bit unfair here, but people seemed so surprised by the discovery of the so-called "dark energy," and I always find *that* mysterious too. Because every cosmology book that I ever knew talks about Einstein's cosmological constant. They solve the equation with the cosmological constant in it.

I've been to conferences where people were saying, *"Well, we don't know what the value is. Maybe it's not there, maybe it's zero, but perhaps by the next conference we will have made a measurement which will have told us what the value of the cosmological constant is. That would be very nice."*

Of course they didn't.

But then when people actually *defined it*—another Nobel Prize—by the wonderful observations of these supernovae stars, and found that, indeed, the universe is accelerating in its expansion, then suddenly this acceleration was mysterious stuff. But it's *already* in all the cosmology books as the cosmological constant.

HB: Do you think that this might have had something to do with the history of the cosmological constant—that Einstein put it in out of a desire to make sure that the universe was static before the observations of the expansion by Hubble? He seemed to have had this attitude of, *Oh my goodness, my equations are showing evidence that the universe is expanding, but we can't have that because the universe isn't expanding. I have to see if I can find a way to do this...* So he added this constant.

And then, of course, it turned out that the universe **was** expanding, and he later recanted and called it his greatest mistake and so forth. Do you think that all of that might have in some ways tarnished the very idea of the cosmological constant?

RP: The whole subject is full of ironies, because not only was he *right* to put it in, but he put it in for the *wrong* reasons. And then he took it out. So yes, it's full of ironies.

But it was there: it couldn't just be forgotten about. *He* might have wanted to forget about it, but the cosmologists didn't. It's in all the books. And that's the part *I* found surprising.

HB: So, it's there, but somehow when it's found to have a non-zero value, it becomes a huge shock.

RP: Yes. It has a positive value, which produces this acceleration—that a positive value produces such an acceleration was, of course, well known. Of course, I think there was a feeling that it shouldn't really be there.

I think that's perhaps what you're really saying: that there was the feeling, *Well, it just messes up the equations a bit.* I had a similar feeling myself, in fact—I'm not going to pretend that I had this insight that it really existed.

But like many other cosmologists, when I could see to put it in, I would put it in. So I knew what the consequences of having the cosmological constant would be to other things I was doing. That was quite clear. It wasn't a puzzle that you could have a term like that.

It was, perhaps, a puzzle to find it *was* there, and for it to be positive with such a small value. I think that's probably another psychological reason for the surprise. If it's going to be there, why is it so tiny?

HB: So, let's discuss that—it's a small diversion, but it seems worth doing so under the circumstances. Just to recap somewhat first, though, there are these two fundamentally different phenomena.

There is dark matter, which first arose from people being unable to make the mathematics work out for the rotation curve of galaxies, recognizing that there should be a lot more matter for the galaxies to rotate in the way that they do, and we don't seem to be able to see it at all. Now the situation is more complicated than just explaining the rotation curves of galaxies, and present estimates show that something like 70% of matter in the universe is actually dark matter.

RP: Yes, most of the actual matter out there.

HB: And then there's the realization that not only is the universe expanding—which had long been recognized—but it's actually *accelerating* in its expansion. Although the mathematical term for this acceleration, the cosmological constant, is not zero but is very, very small.

And related to this there's another issue—that if you look at things from a particle physics perspective, and you try to calculate the cosmological constant from the vacuum energy of particle physics, you get a rather different number.

RP: Indeed. Yes, by what we would call a hundred and twenty orders of magnitude, which is equivalent to a 1 followed by 120 zeros.

It's even worse than that in a sense, because the real answer is infinity. In quantum field theory there are ways of dealing with infinities. There are all sorts of tricks you can do.

It's accepted as one of the most inaccurate calculations done in theoretical physics. From my point of view what it tells you is that it's not the right explanation.

HB: So let's try to move in a more positive direction...

RP: Well, I don't have an *explanation* for this, but let me put it another way: the theory that I'm proposing, Conformal Cyclic Cosmology (CCC), *wouldn't work* without dark matter and so-called dark energy—the cosmological constant.

Perhaps I could show you something that I think would clarify things a bit. This is a picture of the universe.

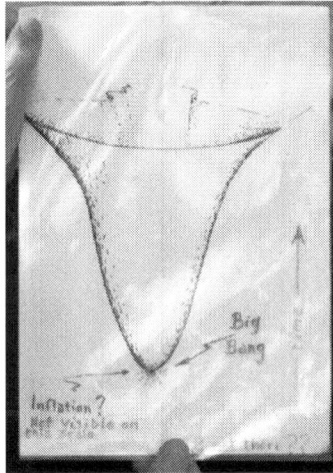

It's a space-time picture, with time going up the vertical axis. That's the way we relativists—as we tend to call ourselves, rather misleadingly— typically view things, with time going up the picture. Horizontal sections through the diagram represent space.

Of course I had to throw away a dimension or two in order to represent 3 dimensions with this one horizontal dimension. But you can see the evolution of the universe: starting with the Big Bang at the bottom and expanding out. The expansion slows down a bit, and then it goes through this accelerated expansion, which is the dark energy that people talk about, or the cosmological constant.

Don't worry too much about the wiggles at the top of the picture at the back—I'm not going to concern myself with whether or not the universe is spatially closed or open—that's not important for this discussion.

So we have the Big Bang at the very beginning, rapid expansion thereafter, then moderate expansion, and then exponential expansion.

Now in modern cosmology books, you will find this thing called cosmic inflation. So you might ask why I haven't put that in the picture?

Well, I don't believe it. I've had a lot of trouble with the idea of inflation for a long time, and I've felt like almost a lone voice in complaining about it. The main reason I have an issue with it is that it doesn't do many of the things it's supposed to do. It does some things, and it's useful in trying to explain some of the details that we see in the CMB, but in my view it doesn't explain the big problem.

HB: So what does it say, first of all? Give us a synopsis of what those who believe in inflation say actually happened.

RP: One of the puzzles, they say, is the spatial uniformity of the universe: *Why is it so uniform?* Another puzzle is, *Why is the microwave background so uniform in temperature?* The temperature over there is very much the same as the temperature way over here. Yet those two points, right back to the Big Bang, were never in causal contact: there was no way of getting a signal from one point to the other in the standard cosmology pictures. So how could it have thermalized to become uniform?

The explanation, according to inflation, is that, first of all, this big expansion smoothed everything out. It might have been irregular at the beginning, but it stretched itself out by this huge factor, and so whatever it was doing at the beginning now it would be very smooth.

When I first heard of this, there were all sorts of things that worried me about it. I thought it was a very ugly idea, and it didn't seem to fit in with the rest of cosmology, but those aren't really scientific reasons.

My principal concern was that it *didn't* actually explain what it was supposed to explain.

And the reason for that is that in my view the bigger problem, which is bigger than *all* these problems, is the **Second Law of Thermodynamics**. The spatial uniformity that inflation is designed to address is *part* of this problem: *How is it that the gravitational degrees of freedom were not activated?* Putting inflation in doesn't solve it.

In fact, I can show you why it doesn't solve it. It's so remarkably simple I can't see why others haven't been copying this idea endlessly.

I'm going to turn the universe upside down. Why? Well, because I'm now going to think of time as going in reverse back towards the Big Bang.

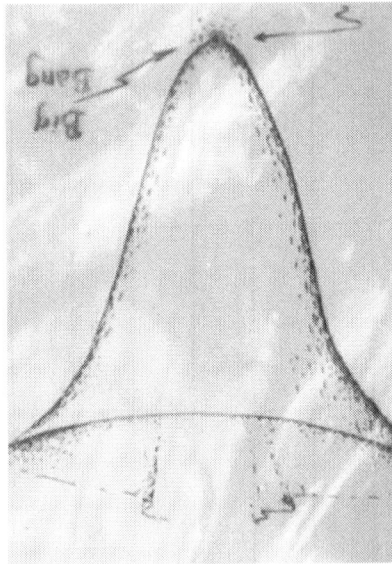

Now in some models the universe collapses and in others, it doesn't—that's not the point: the point is that you can certainly *think* of the universe as collapsing.

So suppose it's a universe that is collapsing and has a Second Law of Thermodynamics in it, so that the entropy is increasing through gravitational clumping. There will thus be irregularities as the universe collapses, and they will increase and will form black holes, and generally make a huge, horrendous mess. Here is that enormous mess.

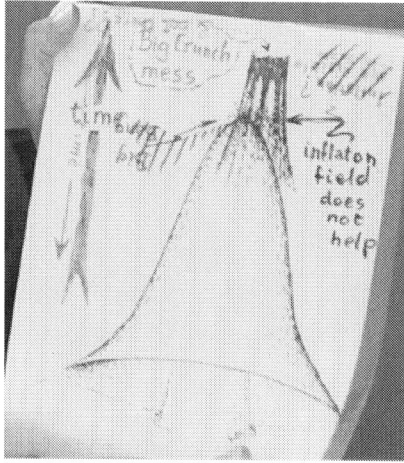

That is, almost certainly, what our universe would do. And when I say "almost certainly" I mean with absolutely fantastic probabilities.

HB: In fact, you've calculated it.

RP: Yes, you can calculate the probability. It's a spectacularly huge likelihood, because when you're talking about entropy there's a logarithm coming in and so on, so the final result is absolutely vastly huge—you can't even write it down with 1's and 0's because the number of 0's is already one of these huge numbers. So let's not worry about the details—the probability is simply enormous.

So our picture becomes a great mess of congealing singularities. I should say that the word 'singularity' is used when your equations go infinite or something goes wrong with them. So infinities come in, the curvatures become infinite here, the densities become infinite there, and you can't do anything. Einstein's equations say, *Whoops—I can't do anything for you.*

HB: I should interject at this point and emphasize to the reader that much of what we understand about the nature of singularities, we actually owe to Roger and his groundbreaking work in the 1960s with Stephen Hawking.

RP: Well, there are also some ironies here that I think are actually quite relevant to our discussion. The original theorem that I had on black holes showed that singularities came about no matter how irregular the collapse was, or whatever kind of matter you had, just so long as the energy densities weren't negative.

And then Stephen picked up on this and applied these techniques to cosmological situations. Then we collaborated on a paper that encompassed most of these results.

But there is a sort of irony there too, because at the time I was thinking, *Why are we limiting ourselves to these simple models of the universe? We could have all sorts of complicated things. That's why you need the singularity theorems.*

But the point is that the universe is not like that.

I remember being in a car near Princeton going to some conference, and in this car was Jim Peebles, one of the world's most distinguished cosmologists.

And I was saying, "*Surely people should have considered all these complicated things that might happen...*" And he replied, "*But it's not like that. The universe is this very uniform state.*"

And I thought, *By God, yes!* That is what started me thinking that this was the **real** puzzle: why is the universe so smooth and uniform instead of such a great big mess?

Almost all of the calculations that people do are in a background of a very smoothed out universe. They put in a little perturbation here or there, but that doesn't really come to terms with the problem.

If we had been in this unbelievable messy situation to begin with then inflation wouldn't do *anything* for us at all. The "unbelievable mess" would have been a state of enormously high entropy—in terms of the gravitational degrees of freedom—and inflation, being a time-reversible dynamical process acting in accordance with the Second Law of Thermodynamics, wouldn't be any use at all: it would just spread out the clumps. So it's really **no** explanation to the question of why our universe is so uniform.

Questions for Discussion:

1. Have you heard of "dark matter" and "dark energy" before reading this chapter? Do you think that science journalists and popular science books on the whole do a good job at conveying to the public what the relevant issues are? (Those with a strong interest in these particular issues are referred to Chapters 6-10 of the Ideas Roadshow conversation **Cosmological Conundrums** *with University of Pennsylvania physicist Justin Khoury.)*

2. Why do you think that Roger's criticisms of cosmic inflation are not more widely shared by the theoretical physics community? (Readers interested in this issue are referred to the Ideas Roadshow conversation **Inflated Expectations: A Cosmological Tale** *with Princeton University physicist Paul Steinhardt.)*

IV. Conformal Geometry

Reinterpreting the end and the beginning

HB: My understanding is that your theory of conformal cyclic cosmology (CCC) is an attempt to solve this muddle, to give some sort of explanation for how we found ourselves at the outset in this smooth, uniform state with very low entropy.

RP: Yes, but I should say that while there is arguably some evidence for the inflationary model in the details of the perturbations you see in the universe, what I'm saying is that that's not really the right way of looking at it. I'm saying that there was a sort of inflation, but it was in what I call the "aeon" that preceded ours.

Now this is an idea first suggested by the Italian physicist and cosmologist Gabriele Veneziano: that there was, perhaps, a scenario a bit like what I'm saying here, where the inflation happened before the Big Bang.

Now, how do you make sense of inflation taking place before the Big Bang?

Here I have a picture of two tricks.

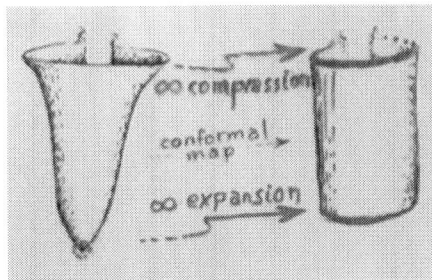

These are mathematical tricks that we've known about for quite some time. I've played around with one of these things for a long time trying to understand how you'd describe gravitational radiation in a nice neat geometrical way. I like to do things in a geometrical way, rather than just flooding myself with equations. If I can see a picture, that will make a lot more sense to me.

Here we have the universe more or less as I had it before. Now I'm going to do two tricks.

If you want to talk about the very remote future—if you want to know how gravitational waves behave, how you measure their energy, and things like that—it's very handy to take infinity and "squash it down". That's a mathematical thing to do, and I want to talk a little bit more about what's being done there. It's what's called a "conformal map".

The whole of future infinity, represented at the top-left, will be "squashed down" into this finite surface—the pink ring-like shape on the at the top of the right-hand side. Of course it's really representing 3 dimensions, and you have to use your imagination there, but the point is that this 3-dimensional top boundary surface on the left now becomes this finite boundary on the right-hand side.

It's just a mathematical trick: we use a conformal map to "squash infinity in". Think about these Escher pictures with angels and devils on a disc—very beautiful pictures. It looks as though it's getting very, very crowded near the edge of the disc, but that's just a map of the whole infinite universe of that particular kind, squashed into the interior of a disc. And that's the same trick that's being used here.

So, although the angels and devils look very small there, *they* don't think they're small. If you ask any angel or devil towards the edge of the disc how big he thinks he is, he'd say, "*Well, I'm just the same size as those ones you think are in the middle*".

HB: But even to us, when they look small, we can see that the angles of the lines between the angels and the devils are the same no matter how big or small the figures seem to us.

RP: That's right. That's the key thing to conformal, it means that *angles* are important, that the angle is the measure of your geometry. We think of geometry as "geo-metry", that means measuring the distance of things, so you get your tape measures or rulers to measure distances or clocks to measure time.

But here you're saying, "*No, the fundamental thing is **angles**.*" In conformal geometry angles are the crucial thing. I'll talk a bit more about how we do this in relativity in a minute.

In this conformal picture, just like the Escher picture, we've squashed infinity down to something nice and finite.

Now if there's a positive cosmological constant, as we seem to have, you can pretty well always do that. There are nice theorems by Helmut Friedrich, who worked on these things to show that this is the sort of thing that you would expect in very general circumstances, to be able to squash this infinity down.

The other part of the trick—at the lower part of each figure in orange—is the opposite sort of thing, which is really what we've been talking about so far: *How do you represent the Big Bang?*

Now here you do the opposite to what we were talking about a moment ago: you stretch it out. Now, this isn't something you can *always* do, but it's a very nice way of describing how special the universe was. My colleague Paul Tod in Oxford formulated it precisely.

I used to talk about the Weyl curvature hypothesis, which is a complicated thing with lots of indices and components that measures the conformal geometry, but Paul's formulation is actually much nicer.

The idea is to say that the measure of conformal geometry goes to zero down here: in other words, that the gravitational degrees of freedom are zero right at the beginning of the cosmological singularity.

But this is a much nicer way of saying it: rather than talking about the conformal tensor, or Weyl tensor, you can use Paul Tod's trick to say that the conformal geometry is smooth at that edge: you can stretch it out and it looks nice and smooth.

So these two tricks give a very nice mathematical way of talking about two things: how do you talk about infinity? And how do you talk about the very special conditions of the Big Bang? It's not saying *why* it's like that, it's just describing, mathematically, what it's like.

And what we see is that, conformally-speaking, the Big Bang is remarkably similar to the other end, to the far future. If you think of the remote future, what would we expect? Well, there'll be a lot of black holes that will be created, and those are the most irregular things: that's where most of the entropy will be.

Yet again irony comes into play here, because here we need to bring in Stephen Hawking and the wonderful thing he did showing that black holes have a temperature. Not only do they have an entropy, which is measured in terms of the surface area of the horizon —Beckenstein originally had that idea but it was made more precise and clearer by Hawking—but he then also introduced the idea that they have a temperature—just for physical consistency.

You don't even need the equations to see this clearly: you can just work with the physical consistency of the Second Law. If you have such a thing as a black hole, it's going to have a huge entropy—you know that because all the degrees of freedom have somehow got lost in the description of the thing—and therefore it has to have a temperature from general thermodynamic considerations.

It's got to be, then—well, not completely cold. How hot is it, you might ask? Well, what are the hottest black holes around? The hottest ones are the smallest ones, which are the results of the collapses of stars a few times more massive than our sun. How hot are they, exactly? Well, think of the coldest temperature ever made on the earth, and it's that sort of ballpark.

It turns out that there are a lot bigger ones around: in the center of our galaxy there is a black hole with a mass of about 4 million times that of the sun. So there is this huge thing in the middle of our galaxy attracting these stars, and they will spiral in and it will swallow them up. Our galaxy is not particularly exceptional: Andromeda has got an even bigger one. We're going to run into it someday, as it happens—we'll come to that in a minute.

But what does Hawking say? He says that even that "supermassive" black hole is not completely cold. It has a temperature much lower than the other ones we've been talking about, because it's much bigger—it's at a ridiculously low temperature. But it *has* a temperature.

The universe will continue expanding, and this microwave background temperature—which is 2.7 degrees above absolute zero—will continue to decrease; and ultimately it will get colder than even the coldest black holes around.

At that point the black holes become the hottest things around and they will start to radiate. And according to $E=mc^2$, the mass gets taken away, so that the mass of that hole shrinks down very, very slowly.

No matter how big they are to begin with, they will ultimately shrink down and down to this tiny Planck scale and go off with a pop. I call it a pop, but it depends on what time-scales you're thinking about: a bang might be reasonable because it's like a nuclear explosion, but at the very end is a pop—it has a fairly sizable bang before that, but this is small beer on the cosmological scale.

HB: But at any rate, all of these black holes will go away and we will be left with just radiation.

RP: That's right.

I used to think that one shouldn't use emotional arguments in physics. Or should you? Well, you see, this is a bit of an emotional argument: I call this "the very boring era". It's very boring sitting around waiting for these things to go off—because for the biggest ones we know, which are something like 10^{10} solar masses or something like that, we estimate it will take them something like 10^{100} years to radiate away—a simply enormous period of time. So that's pretty boring to be waiting around for that to happen.

But then I thought, *Who's going to be bored by it?* Well, not us. The only things around at that point will be things like photons, and it's very hard to bore a photon. And to demonstrate what I mean by that, I'll need to talk a little bit about light cones.

Questions for Discussion:

*1. Have you seen Escher's **Angels and Demons** woodcut? Does this chapter make you more interested in looking at other Escher works? Those interested in this topic may also appreciate that Roger and his father corresponded with Escher in the 1950s—and, in developing the "Penrose tribar" they strongly influenced some of Escher's other works, such as **Ascending and Descending**.*

---

2. Why do you think that black holes need to wait until they are warmer than the cosmic microwave background before they begin to radiate?

V. CCC: The Basic Idea

Matching beginnings and ends, conformally

RP: Before I do that, let me just start by showing you the general picture of Conformal Cyclic Cosmology.

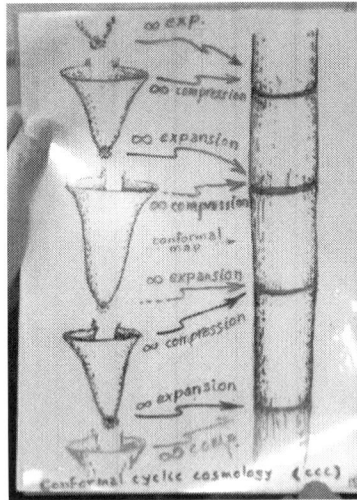

Conformal cyclic cosmology (ccc)

If we look at the conformal pictures on the right-hand side of this transparency, then by having a smooth boundary between the beginning of one "aeon"—which, by hypothesis involves a very low entropy due to gravitation—and that of the end of the previous "aeon"—just due to the way the equations dictate—then we might at least imagine that this is part of an infinite continuing succession.

HB: So there's a recognition that if you look at things appropriately, by understanding black holes radiating and so forth, then the very, very far future of our universe is very similar to our state right at the Big Bang, which enables you to consider making this big conformal chain.

RP: Yes. And 'similar' is a good word, because when you talk about geometry, we sometimes talk of "similar triangles"—triangles of different sizes but the same shape. That's a *conformal* geometry concept: we're not interested in size, what we care about are the angles. If the angles are the same for two triangles, they're similar triangles—they're conformally the same.

Now so far this is just a conformal picture. The measures of distance or time on the left-hand side, which might look infinite, is all squashed down into that finite region on the conformal picture on the right-hand side by that mathematical trick I showed you a moment ago.

Now you might well ask if that makes any sense *physically*? Well, I'll come to that. But, I'm just doing geometry here. Each part of the chain is a nice representation of what I'm now calling our "aeon". And there is, presumably, an infinite succession of these aeons. That's the basic picture.

Now suppose our universe physically collapsed instead of expanding to infinity, and it did that before the black holes went away. In that case, you'd have a great mess, and the corresponding conformal picture of the end of that aeon wouldn't glue on very well to the next one.

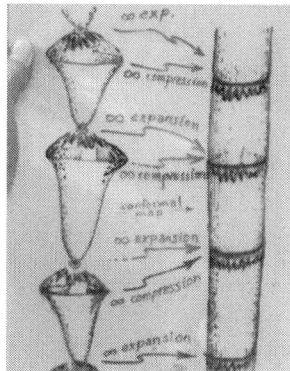

HB: So we're *not* talking about a model where the universe began with a Big Bang and then collapsed into something which then somehow "bounced" into something else. That's not what we're talking about here.

RP: That's right—and you can see clearly why that's not what we're talking about, because if things become so irregular that we produce a big, clumping mess, then this chain of conformal maps won't hang together properly.

HB: So in your model, the end of one aeon is necessarily very similar to the beginning of the next.

RP: Yes, it is remarkably similar, if you think about it. The beginning of the universe is this very smoothed-out, thermal, state, which is not at odds with our current picture of what will ultimately happen with the universe.

Questions for Discussion:

1. Is there any reason to necessarily believe that the chain of successive aeons is infinite? Might it be the case that, for some reason, there was one "Super Big Bang" that began the process and one eventual final end state? What conceptual difficulties would be naturally associated with that view?

2. What does Roger's analysis imply about other cyclic models of the universe that involve a "bounce"? How might they address the entropy initial conditions problem that motivated Roger to begin with?

VI. Light Cones

Photons vs. massive particles

HB: So just to summarize: the idea is that we start off with our Big Bang, and we say to ourselves, *"Isn't it peculiar that, coming right out of the Big Bang, we happen to find ourselves in this extremely low entropy state where everything is so smoothly and uniformly distributed?"* But we file that in the back of our minds and we say, *"Well, let's just watch and see what happens"*.

So we watch and see what happens: gravity "turns on" as it were, we get galaxies, stars, and eventually black holes; and these black holes get bigger and eat other black holes and eventually— after an exceptionally long period of time—they radiate away into photons.

And we notice that at this very much later point, we can look at things in a conformal way and note that it actually seems strikingly to the situation right after the Big Bang.

So we can identify this very, very far future through a conformal mapping with the Big Bang, prompting us to imagine a scenario where we might have a cyclic nature of aeon after aeon.

RP: That's right.

HB: But a moment ago you were talking just now about using conformal maps as a mathematical trick, effectively saying, *"If I look at things in this particular way and don't care about things like distances and only worry about angles and things analogous to similar triangles, I can make this identification between the physical universe and this conformal picture."*

But in the real world we *do* care about things like distances. So how can you convince me that this conformal picture actually matches up with the important things we care about in the real world?

RP: Well, let me talk about light cones. If I'm talking to my colleagues I would call them "null cones", but never mind—let's call them "light cones" for now.

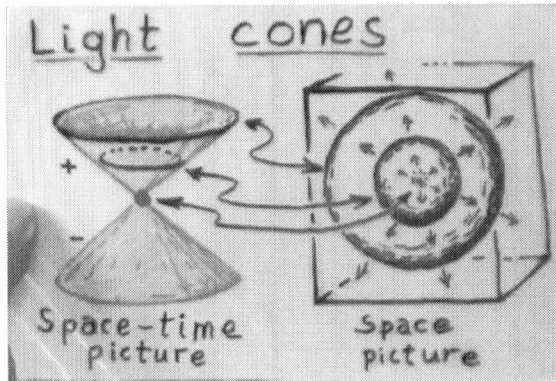

On the left we have a space-time picture with time increasing upwards, whereas on the right we have the corresponding spatial picture. Now you must think that at every point of our universe, every point of space-time, there's one of these light cones imagined to be there, present in the locality of that point.

It's telling us what light would do there: if you had a flash of light at that point, its history would spread out along this cone. This initial flash is represented by the red dot in the middle of the space-time picture, and the center of the ball in the spatial picture on the right-hand side. I've shown the corresponding pictures as time increases in both the space-time and spatial pictures. But the key point to remember is that at every point in space-time there's one of these things—it's not necessarily telling you what light is actually doing at any given moment, it's telling you what it *would* do, both in the past and in the future.

Beneath the red dot on the light cone at the top left is the past light cone where you have converging light on to that point in space-time from the past and above the red dot is the future light cone, to give you a clear time orientation for the whole picture.

Now, Einstein's special theory of relativity uses something called Minkowski space—it's a nice, four-dimensional geometry—and in this four-dimensional space you've got one of these cones at every point.

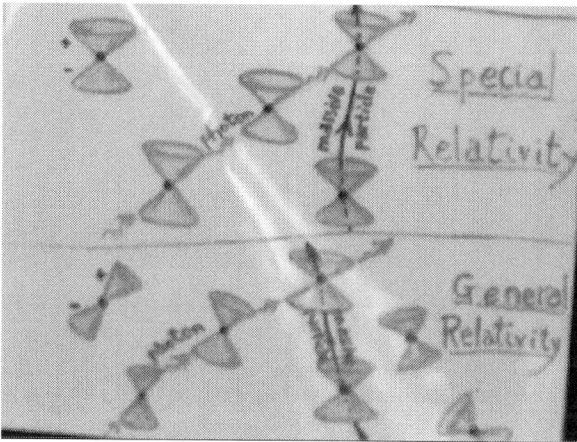

Now this is telling you a very basic thing about physics: it's telling you how particles behave—they're not allowed to go faster than light. On the right of the picture you can see a black line going up the picture. That represents the "world line" of that particle—it's a history of the particle, where its speed is given by the slope of that line.

That slope must always be less than the cone here, so that the world line of a massive particle—a so-called "timelike line"—is always contained within the cone, whereas a massless particle like a photon moves along the edges of the cone. So the photons travel *along* the cones, while massive particles always travel *within* the cones.

In general relativity the picture is more or less the same, just that the cones are not uniformly arranged—they can wiggle this way or that way—they do that within black holes, for example: they're quite complicated in the way they behave. But it's still the same rules: photons along the cones, massive particles within the cones, with each cone being split into a future and past cone.

So that is the way we think of space-time. Now, what about the metric? How do we measure distances and times?

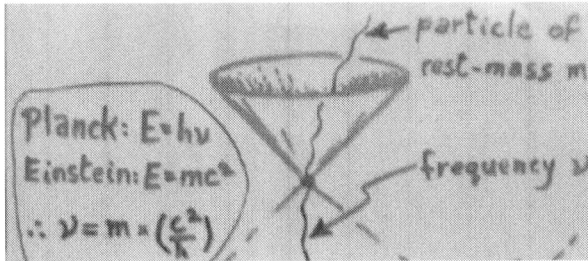

Well, we do have extremely accurate clocks now, such as nuclear and atomic clocks, which ultimately depend on very basic physical principles—Einstein's $E=mc^2$ and Planck's $E=h\nu$—which tell you that any massive particle is a little clock on itself, a clock with a frequency which is given in proportion to its mass.

So we have a particle that we think of as oscillating with a quantum oscillation of very high frequency that is given purely by its mass. In nature, then, we have incredibly accurate clocks; and we can harness this in our atomic and nuclear clocks with enormous precision.

Now, I'd like to examine the clocks associated with two different particles moving inside my light cone at different speeds, where I can represent the ticking of each of these clocks by these bowl-shaped surfaces on my diagram inside the light cone.

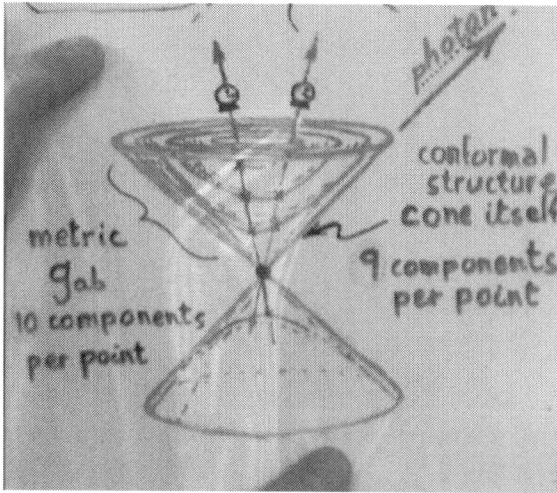

The point I want to make is that the structure of space-time depends on two things. One is the structure of the light cones that we saw a moment ago and the other is how crowded these bowl-shaped surfaces are.

In Einstein's theory, there's this thing called g_{ab}, which we refer to as the metric tensor, that carries ten pieces of information at each point. Nine of them—more precisely, the nine independent ratios of those ten numbers—tell you where the light cone is. And the tenth number gives you the scaling: how crowded these bowl-shaped surfaces are inside the light cone. Together they give you the full metric, the full space-time picture.

Questions for Discussion:

1. Given that the concepts of light cones and Minkowski space were developed several years after Einstein first formulated special relativity, to what extent does this demonstrate how a geometric formulation can significantly deepen our conceptual understanding? Is is possible to develop the full appreciation of the relevant concepts without such a picture?

2. To what extent is it possible to regard the claim that "nothing can go faster than light" as a consequence of a more general insight denying non-instantaneous communication between two points, rather than a seemingly arbitrary declaration of a "universal speed limit"?

VII. Massive Headaches

And how we might make them disappear

HB: You were talking before about how we can use a particle as a clock. But what if my particle doesn't have a rest mass? What if my particle is a photon, for example, so it's moving at the speed of light?

RP: Yes, well, that's precisely the point: then you *can't* use it as a clock. If you were a photon, you wouldn't feel the passage of time at all—there's zero time between any two moments on your world line. Infinity, if you like, is just like now: there's no passage of time whatsoever. Another way to put it is that, for a photon, everything happens at once.

Remember a moment ago, when I was talking about this very boring final expansion era of the universe? Well, the photons in that universe don't experience the passage of time at all, which is why they couldn't be bored at all. Infinity is "just there", as far as they're concerned.

How do you describe electromagnetism and electromagnetic waves? Well, we have these incredible equations due to Maxwell, the greatest physicist between Newton and Einstein.

Maxwell's equations were built on the ideas of Faraday, who had done seminal experiments, and had this idea that maybe the fields had a reality. Maxwell developed the equations to describe that reality that from that moment on had to be adjoined to the physics of particles: these fields filling space.

And these equations of Maxwell, which can be regarded as the prototype of all the physics equations we have now, have the particular property of being conformally invariant: they don't need that tenth number that I was talking about before. They only need the nine numbers—they only need to know where the light cones are.

You'll recall that those nine numbers don't represent the whole of Einstein's geometry—you need the tenth number too, which tells you about the extent of the crowding of these bowl-shaped surfaces within the light cones.

Some parts of physics need the full metric geometry, while some parts get away with just conformal geometry. Maxwell's equations, as well as other equations that govern other forces of nature such as the so-called strong and weak forces, just get away with conformal geometry.

HB: And when you were creating this correspondence between the very late universe with the very early universe, you were also using conformal geometry.

RP: Yes, we're just using that. If I just had the light cones, these pictures I showed you of mapping successive aeons through conformal transformations would make perfect sense: the cones will carry along completely happily as you go through successive aeons. That kind of geometry—the conformal space-time geometry—just uses the light cones.

The only time, as far we know, that you need the full metric geometry is when mass comes in. When we describe things like the strong and weak forces for particles with a given mass, the equations get more complicated, and the conformal invariance is lost.

And on the other side of things, as it were, mass is also, of course, the source of gravity. So Einstein's general theory of relativity does need that other parameter, the tenth number, as do the parts of physics that involve the masses of particles.

HB: So there's a big difference between these two pictures, between the conformal picture and the full metric picture, depending on whether or not we have massive particles around.

It seems to me that this gets back to this idea of whether or not we can create clocks with our particles. If you're a photon, as you say, there is no other moment: you're sort of slamming into infinity, as it were.

RP: Yes. That's right.

HB: But it's very different if you're a massive particle. So, can you talk about that in terms of what's happening at the very end of the universe? How is all this relevant to that?

RP: Yes, well, you might well simply ask, *How is this actually relevant to physics?* Because there **is** mass in physics.

But before I get to the very end of the universe, let me talk for a moment about the other end—the very beginning. Because there, aspects of my story turn out to be not so unconventional.

Let's talk for a moment about the Higgs boson. The Higgs is a mechanism for the origin of mass. It has its own mass, which means that there is a certain temperature in the early universe at which the Higgs kicks in.

And if you go back before that time, the Higgs is effectively massless, as are all the other particles—the energy in individual particles is given by their motion, if you like, rather than their rest mass. The rest mass becomes more and more irrelevant the hotter it gets.

There are detailed issues to address here, but the basic point is that the closer you get to the Big Bang the more irrelevant mass becomes, and you really *are* dealing with massless entities.

Since everything is massless in the Big Bang, in a certain sense this conformal picture I was talking about before of successive aeons is not simply mathematics—which is to say that the things that are around really don't know that there is a Big Bang at all. They are not particularly concerned with that boundary between one aeon ending and another one beginning.

People tend to think of the Big Bang as a point. They say, "*Our universe began as something terribly small, smaller than a pinhead*", or whatever.

But that's not the right way of looking at it, because if you're looking at it from the perspective of conformal geometry—as opposed to metric geometry—which is the relevant geometry because rest mass is not yet relevant—it's stretched out, so the Big Bang was a huge three-dimensional space.

And the sorts of problems that people have of correlating different points in the early universe arise from this confusion, you see.

As I mentioned earlier, a key question that inflation was created to solve was, *How is it possible for the temperature in the CMB to be almost the same in two different places that were never in causal contact?*

Well, in the conformal picture—and this is not my theory per se, this is just standard geometry—they're *not* close to each other. There's no signal that can get from one to the other. This is making that more explicit: in that conformal geometry, they're simply *not* in causal contact. They're in separate places.

So if there is some correlation between two points that couldn't have been in causal contact with each other in the very early universe, it was because of something that happened *before*, at an earlier phase of a previous aeon, where these points *were* in causal contact with each other.

Addressing this problem of correlations was one of the main problems that inflation was meant to deal with. Well, inflation does it for you—albeit rather artificially, in my view. But in my picture it does it automatically, by saying that they became correlated in an earlier aeon.

HB: But it seems to me that if we look at the other side of things at the boundary between aeons—that is, if we go to the very, very far future of one aeon—in order to have this conformal identification, we clearly can't have very much mass lying around.

RP: Yes, absolutely.

HB: OK, so let me try to focus on things now from a physics perspective rather than a geometric perspective. We've got all sorts of massive stuff in the universe now. And in order to get to the final state that it needs to be in for this model to work, where I can map the end state of one aeon to the beginning state of another, I have to somehow get rid of all this mass. So what happens to the mass that we see around us? Well, most of it goes into black holes, I guess, and then radiates away from black holes through Hawking radiation and turns into massless photons.

But presumably I must have *some* mass that's around somewhere that doesn't fall into a black hole. How can I make that geometric equivalence that you keep referring to—that conformal mapping—that implies that I must have only massless particles around? What happens to normal electrons and other things?

RP: You hit the nail on the head of a very basic point—which is, as people have sometimes said to me, the weakest point in the whole theory. And I agree with that. You could say, "*Well, most of the particles are photons by far so let's not worry about the odd electron*", but I'm not happy with that. That electron is a clock, and so it needs to know the metric.

HB: And then the conformal invariance is broken.

RP: That's right. *You might think, Well, protons might decay,* but even if that might somehow be the case, the electrons have got nowhere to go. One can imagine there being a less massive, charged particle that they end up as somehow, but the whole point is that it must be *truly* massless for this scheme to work. Are there massless charged particles?

Well, it's possible that there are massless neutrinos, but they've got no charge, of course. As it happens, neutrinos are known to have a mass—that is, at least 2 of the 3 are known to have a mass—and probably all 3 have masses.

But even so, neutrinos therefore aren't going to be any help to us here because they don't have a charge, and that's no help here because we'd have to violate charge conservation, which is going a bit too far—even for me. I'm not prepared to do that.

But let's look closer at mass.

I've been talking about the wonders of rest mass and how it gives us our incredibly accurate clocks and all that, but our theories of particle physics as they currently exist depend on something called the representations of the Poincaré group.

It's not so important what that means in detail, but what it basically means is that we're doing physics without a cosmological constant: with a cosmological constant, you find some little problems involving the rest mass—that it's not quite as constant or invariant as it should be. Because in ordinary physics without the cosmological constant there's something called a Casimir operator, which is related to the rest mass. Again, we don't need to worry about the details here, but the point is that normally we think of mass as the fundamental thing that should be constant for a stable particle. But when you have a cosmological constant, it's not so clear.

So I'm hypothesizing that there is a sort of anti-Higgs mechanism going on, which kicks in, in the very, very late universe.

As we've said, the general picture is that in the very early universe, very close to the Big Bang, the Higgs mechanism kicks in and provides mass to particles.

Now, what I'm proposing is to imagine that mass faded out on the same sort of scale on the other side as in this picture, which is to say *extremely* far away in the future, long after all the black holes have evaporated away—when 10^{100} years seem like almost nothing.

I'm not saying when that *is*, exactly: I'm just saying that you need it to make this theory make sense.

Clearly there is no massless charged particle around now which the electron could decay into, because that doesn't work, as I'm told on good authority by particle physicists. That would represent a severe contradiction with current experiments.

However, there's nothing against having this gradual fade-out of mass whereby all particles would gradually become massless in the asymptotic limit. I'm not saying that they ever become completely massless, but that the mass fades away sufficiently so that this conformal picture makes sense.

That's a hypothesis which doesn't seem to me to be so outrageous, because it's like saying that the Higgs not only has a certain temperature when it kicks in, it also has a certain— very low— temperature when it "kicks out," if you like. And the equations seem to require something like that.

HB: It's a necessary hypothesis in order to get this matching between the very late universe of one aeon and the very early universe of another.

RP: Yes. So just at the crossover, the electrons would have become massless things that could get through to the other side.

Questions for Discussion:

1. What does Roger mean exactly, when he says, while discussing the problems of mass associated with an electron, that "neutrinos aren't going to be any help here because we'd have to violate charge conservation"?

2. To what extent could one claim that a posited "anti-Higgs" mechanism that "kicks out" at the very end of an aeon is justified by a type of "symmetry argument" to one that "kicks in" at the very beginning of an aeon? To what extent might one claim that the entire theoretical framework of CCC is strongly motivated by symmetric considerations?

VIII. Necessary Conditions

Dark energy, dark matter and information loss in black holes

HB: My understanding is that, unlike standard cosmological pictures where people say, "*This is our picture of the universe, but there's also this dark matter and dark energy that don't seem to quite fit with the other things*," from your perspective both dark matter and dark energy are actually integral to conformal cyclic cosmology.

RP: Yes. Perhaps I could talk about dark energy first, because it's easier in a certain sense. Before I really believed in the theory—I was a bit slow in picking up on this issue, I have to say, because a lot of the work I had done previously assumed that the cosmological constant was zero, which is in some respects nicer, but in other respects not so nice.

And one of the differences between the two scenarios is that if there is a positive cosmological constant, then the boundary at the far, far future is space-like. Now that is *crucial*, because we know that the initial boundary at the Big Bang singularity is also space like, so if you're going to be able to glue them together they both have to be of a space-like character, cutting across the line cones.

And so that realization—acceptance, really, throughout the scientific community—of the positive value of the cosmological constant was one of the key ingredients that led me to say, "*Ah! So they fit!*" If you don't have a positive cosmological constant, you just can't make this picture work. So that was a key initial motivation.

Then, later on, I started working out equations trying to make the scaling work as we transition from one aeon to the next. You've got a conformal factor, you see, which in a certain sense turns upside down: the "squashing in" of infinity becomes the "stretching out" of the Big Bang, meaning that you have a conformal factor which turns into its reciprocal, and you have to make sense of that.

And in doing so, you find that you're *forced* to create a new material each time you do so. Roughly speaking, in order to express the Einstein equations in this conformal description, you need to introduce a field that keeps track of the scaling.

That is, from the basic conformal geometry, you need a field that enables you to get back to the full metric of Einstein's general relativity. Now at this point that field is not really there. It's what I call a phantom field. You can work out all the equations and how it behaves under the conformal maps and all that, but it's not a physical field. It doesn't have its own degrees of freedom: it's not there.

But when you turn the conformal factor upside down, you find it's got in there with a vengeance: making by far the biggest contribution to the matter in the new full Einsteinian metric that you have, very much in addition to all the photons and massless things that got through from one aeon to the next. And this is entirely new.

Which leads me to say, "*Well, if it's anything new, it's got to be the dark matter*". It's the major contribution to the material in the universe. It's initially massless—that's wrong from the point of view of what we know about dark matter—"

HB: But presumably it has its own Higgs field or something like that.

RP: Exactly.

In short, there are three things that are needed for this scheme for consistency reasons.

1. A positive cosmological constant—dark energy.

2. Creation of dark matter, vast amounts of new material of a scalar type—a new scalar field that appears.

3. You need rest mass to come in. Again, the equations tell you— and this came as a bit of a surprise to me—that because you've turned the conformal factor upside down, you find that rest mass has to creep back in again. It just isn't consistent without that.

So to my mind, this creeping back of rest mass must tie in to the Higgs idea somehow. I've been trying to work it out in detail— not yet with any huge success—but it seems to me that there are very interesting opportunities to explore there.

These three things all come in because the equations are inconsistent without them. There's no explanation of why any particular values have to be this or that, but without these three separate concepts the whole scheme wouldn't work.

Perhaps there are readers who have, quite correctly, a certain nagging feeling about all of this. After all, didn't I start off this entire discussion by talking about the Second Law of Thermodynamics?

And if you look at my cyclic picture, you might well worry: doesn't the entropy always seem to be going up from one aeon to the next? How can I possibly have a repeating model like this?

Well, I've worried about this for a while—not quite correctly I think. I think it's correct now, but it's one of the points in the argument where I'll encounter a lot of opposition, I can see that.

So let's start with this question: *Where is the entropy now, even? Where is most of the entropy in the universe?*

HB: Black holes, I guess.

RP: Indeed, it's black holes. People used to think that the CMB has a lot of entropy in there, which is true, but it's absolutely chicken feed compared to what's in black holes. Now, what's going to happen to those black holes? Well, as Hawking showed us, they will evaporate away.

But here comes the contentious issue. Initially when Hawking put forward these ideas, he argued that the information that was swallowed by these black holes is lost. And this represents a violation of one of the basic quantum mechanical principles called unitarity. Hawking was prepared to generalize quantum mechanics in order to incorporate this loss of information.

Many years later he changed his mind, and went along with most other physicists in saying that, somehow, the information escapes and is regained.

HB: He famously lost a bet with John Preskill about it.

RP: Right. He was even prepared to lose a bet. Meanwhile there I was, cringing, saying, *"Don't concede! You were right the first time!"*

I think that information **is** lost in black holes. I think the argument he used the first time was very powerful and impressive. You'd draw these conformal diagrams, and you could see that the information had just gone straight into that singularity and was gone.

Now this, of course, goes against certain other areas of physics but some of these areas I'm not so unhappy going against because I don't altogether believe them anyway.

This is another story, but in the way that people often think of quantum mechanics in terms of unitary evolution, then information can't be lost. And that's right. But whenever you make a measurement in quantum mechanics, you say to yourself that this thing is violated all the time. Anyway, I don't want to go into those details right now, but merely explaining why, psychologically, I was quite happy with the idea of information loss in black holes.

HB: But let's back up a moment. If you're saying that information is lost, isn't that a violation of the Second Law of Thermodynamics?

RP: No, there's a subtlety you see. John Wheeler used to use a nice term, "transcended", which I think is very appropriate here: the Second Law isn't violated, it's transcended.

The general principles of thermodynamics so evidently drive this entire discussion: the black holes have a huge entropy due the fact that the number of degrees of freedom needed to describe a black hole is tiny compared to the ones that went to make it; the temperature of a black hole has to exist for general thermodynamic reasons; the eventual evaporation—all of these things are driven by these general thermodynamic reasons. You don't even have to know the details of all of it. The whole story unfolds from very powerful principles coming from the Second Law and the general principles of thermodynamics.

To discuss the matter of information loss in more detail, then you really have to know how to define entropy. I've been talking about it by waving my hands around and saying randomness and so on.

But the real breakthrough here came from Boltzmann. He had this definition of entropy that works in very general circumstances. To make this very clear I'd have to talk about phase space, but the basic idea is to evaluate the entropy of the system based on its degrees of freedom and how, precisely, they're activated.

Now in a black hole, I'm saying that it swallows these degrees of freedom—the information *is* lost—so you don't care about them anymore.

In order to go through this properly you need to go through Boltzmann's formula here, but the main thing to note for our purposes is that this formula has a logarithm in it.

And that means that if you have two quite separate systems with two different numbers of ways each system can be created then to find the number of ways the total system can be created you simply multiply those two numbers.

Now to find the entropy of any system, you take the logarithm of the number of ways it can be made. And since the logarithm of a product of two numbers is equal to the sum of the logarithms, that means that the Boltzmann entropy of the combined system is equal to the sum of the Boltzmann entropy of each subsystem.

In other words, thanks to Boltzmann's beautiful formula with this logarithm in it, entropy becomes additive.

Now, suppose I'm sitting in my lab—I don't sit in a lab normally, I must admit, but pretend for a moment that I would be a good, professional experimental physicist sitting in my lab—and I'm interested in the entropy of the whole room I'm in—imagine it being isolated from the outside. Now outside there's that whole galaxy out there. If I want to know the entire entropy of the system, I take the entropy in this room plus the entropy of the galaxy. There might be a black hole in the center of that galaxy, but I don't care: it's just an additive constant to my entropy. I do my entropy here, and I might add in the one for the black hole but I might not. I probably wouldn't.

Now suppose that black hole disappears, and its degrees of freedom disappear. Well, I don't notice it. The measure of entropy in my lab isn't affected by that black hole. The Second Law as operating outside that black hole works perfectly well.

So the Second Law, in its normal operation in the universe doesn't care if I've suddenly changed the rules by forgetting all those degrees of freedom in the black hole.

That's what I mean by transcended. The Second Law goes its merry way, it works perfectly in every specific instance, there's no violation of it. It's just that I've changed what I *mean* by entropy.

The Second Law always works, but it's because, subtly, the definition of what you meant by entropy has changed. We're just saying, *"Now I don't care about the ones which have got swallowed by the black hole."*

HB: So in order for CCC to work properly, in order to ensure that entropy doesn't improperly "build up" from aeon to successive aeon, it's necessary that information does get lost in a black hole. Which means that you're reopening this debate that seems to have been closed.

RP: I know, but they closed it the wrong way. Indeed, the information *is* destroyed—that's what I'm claiming—and that's the whole way that CCC's framework works. It's because that information is of no interest to anybody anymore, and therefore I don't use it in my measurement of entropy.

If I wanted to keep it in, every time I go through these aeons, it would be going forever upwards. But each time you're throwing it away. It's because you're allowed to throw away those degrees of freedom that you can have a consistent Second Law picture.

Questions for Discussion:

1. How might the public reception to CCC have been different if Roger had predicted a positive cosmological constant as a result of his theoretical model before it was discovered?

2. To what extent does Roger's argument (or John Wheeler's notions of "transcendence") about the need for reformulating our definition of entropy by accounting for information loss in black holes demonstrate an inherent amount of arbitrariness associated with any "law of nature"?

3. If information is not lost in black holes, then how, precisely, might it be "preserved" in the radiation that black holes emit as they disappear?

IX. Cosmic Imprints?

Searching for evidence of past aeons

HB: Is there any experimental evidence for your theory? Are there controversies—or at least differences in opinion—associated with your search for experimental confirmation? What are the specific predictions that CCC makes and why?

RP: Well, I think there is at least one clear prediction—in fact there are probably lots, and we will likely see more as the theory gets more thoroughly worked out. So far it's been a relatively small number of people involved: myself, Paul Tod—who's done a fair amount of work on the equations and influenced them a lot—and my colleague from Armenia, Vahe Gurzadyan, who's doing much of the analysis of the data.

My hope is that there will be interest from other quarters. I think graduate students feel a bit nervous working on this sort of thing because it's a little different from what most other people have been doing.

So let me talk about what I think of as the major observational aspect. There are lots of things that one could argue one way or the other, but let's begin by asking, *"What is the most violent thing in the aeon prior to ours that could have conceivably occurred and left some sort of imprint that we might now be able to see?"*

We do have these enormous, so-called supermassive black holes around—there is something like 4-million-solar-mass black holes in our galaxy. We're heading towards a collision with Andromeda, which is a member of our Local Group of galaxies, and apparently contains a supermassive black hole that is even slightly bigger than ours.

Now when this collision happens—which I should stress isn't for a long time yet—it's quite possible that the black holes might capture each other. They might not, they might miss: it depends on how head-on the collision is. But it's quite possible that they'll capture each other and go into a sort of orbit and spiral around each other, emitting gravitational radiation, the gravitational version of light.

Light is electromagnetic waves, and there are gravitational waves as well, which are extraordinarily weak. The most significant source, in terms of intensity, of gravitational radiation is probably these encounters between black holes. There are ongoing experiments trying to detect gravitational radiation that, within the next decade, will probably directly detect this gravitational radiation, perhaps from collisions between black holes (*Editor's note: gravitational waves were first directly detected by the Laser Interferometer Gravitational-Wave Observatory— LIGO—in September 2015*).

Now when these two super-massive black holes capture each other and congeal, there will be a burst of gravitational wave energy carrying away about a few percent of the total mass energy of those objects.

This is happening, presumably, throughout our aeon. But it should also have happened in the previous aeon. Here is another space-time picture describing the situation, with time moving upwards up the page:

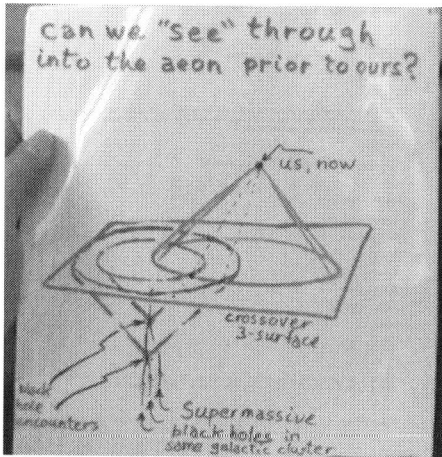

Can we "see" through into the aeon prior to ours?

us, now

crossover 3-surface

black hole encounters

Supermassive black holes in some galactic cluster

The rectangular plane represents the crossover surface between aeons (which you must remember is really a 3-dimensional surface), and above that is us now, looking back through our past light cone to a surface just after the Big Bang to the microwave background.

Meanwhile, prior to all that (following the purple light cones), we are envisioning that, in this previous aeon, a couple of galaxies with super-massive black holes have met and produced a sizable amount of gravitational radiation. In fact, depending on the size of the galactic cluster and the corresponding black holes, this congealing of super-massive black holes might happen several times, resulting in several large bursts of energy.

Now according to the equations of CCC, the energy of the associated gravitational waves will convert itself to energy in the initial dark matter. That comes about from how the conformal factors work: the gravitational degrees of freedom convert themselves into this dark matter in its initial form (which is initially massless—it's a sort of wave in the dark matter), and as the dark matter acquires mass, this wave slows down and becomes a slight movement in this "cold dark matter," as people say—it's like a fluid without much motion in it.

But there will still be a remnant of this impulse in this wave, which is causing the dark matter fluid to move out in the direction of that wave.

And depending on the geometry, we would see some circles in the CMB as associated ripples, if you like, of these waves. In three dimensions, they will be spheres.

And as we look back over time through our light cone to the Big Bang, that will intersect these other spheres producing a circular cross-section in that 3-dimensional space. And that circle is what we will see.

If the motion of the previous gravitational radiation is in our direction, it will result in a slight warming of the CMB. It's a Doppler effect, really: for the relatively far sources—well before the aeon crossover—it's moving towards us, and the resulting impact on the dark matter fluid will be slightly in our direction, resulting in a slightly warmer ring in the CMB. Meanwhile, for the relatively close sources—black hole collisions closer to the aeon crossover—this wave is moving away from us, resulting in a slight cooling of the CMB.

HB: So if this is correct, we should be able to observe these warmer or cooler rings in the CMB. We should be able to measure them.

RP: Well, I did approach people at Princeton and asked them if anybody had seen an effect like this. I talked to David Spergel, who's an expert on this, and I asked if anybody had ever seen them. He told me that nobody had ever looked. So one of his postdocs did some analysis that didn't reveal anything specific. But as it turns out, I think that the way they were doing it might not have been the most promising.

Because of these potential multiple occurrences I was just speaking about—that these encounters of super-massive black holes will happen more than once—this means that these rings might well occur in families of concentric rings.

The initial image I had was something like a pond with drops of water coming down and producing rings in different places. But after a while you can't see these rings because they interfere and produce a great mess of ripples. But if you take these multiple occurrences into account, the appropriate image becomes more like rain dropping off the corner of a roof or something and hits the same spot in the water several times, which produces concentric rings.

HB: Are there issues with sensitivity to be able to measure these things accurately enough? Could it be the case that we perhaps don't have the right technique somehow to be able to sufficiently process this, or be sensitive enough to it, or even know precisely what to be looking for to start with? In short, when do we say, *"Well, we've looked long and hard at the data and this just isn't there, and therefore this theory is incorrect?"* as opposed to, *"It's too early to make any definitive judgements at this point because we don't have sensitive enough equipment, or can't be sure we're looking properly"*, or whatever?

RP: Well, clearly it needs some very detailed work. Now, I can't do all that myself, obviously. I need other people to help me with all sorts of special expertise in different areas: galactic dynamics, to determine how long galaxies will typically encounter each other and how much energy will be produced; detailed astrophysical analysis of the specific forms of energy released through black hole collisions...

HB: So you need all this to know what, precisely, to look for in the CMB data.

RP: Yes, that's right.
And you also need to know the detailed interaction of that energy at the crossover when it hits the new dark matter material—incidentally, that brings up another point I perhaps should have made before: the dark matter has to decay too,

otherwise it would mount up from aeon to aeon, much like my concern of mounting entropy I mentioned before.

This is different than our problem with the electrons we spoke of earlier: it's not just that its mass has to go away, *it* has to go away. Fortunately, there does seem to be some evidence that this sort of thing occurs: I'm not an expert on this, but I gather that our current understanding is that in the early universe there was considerably more dark matter than there is now, so there's reasonable reason to believe that it can decay. And I've got quite a bit of leeway, because I've got the whole rest of the aeon to play with until it goes away. So there's lots of time.

But that's what I need: for the dark matter to pick up these degrees of freedom, a long time before it decays away, and give a kick to the initial material in such a way that the ensuing radiation will incorporate this motion. But to figure out all the details precisely needs a lot of people to do precise and sophisticated calculations.

I have a simple answer, which is begging the question somewhat: if CCC is right, then it seems to me that the major effect in the temperature variations has to be from this process. So you should just about see it.

Now I recognize that this is great optimism on my part, since the first analysis didn't seem to reveal anything. To my mind there were some indications that there was something there, but it wasn't enough to convince anybody else.

But when I mentioned this to Vahe Gurzadyan, an Armenian colleague I've known for some time, he surprised me by showing considerable interest. At the time, I was interested in the potential distortion of these rings through gravitational lensing effects, so I consulted Vahe, because I knew he had ways of analyzing that sort of thing. But to my surprise he became generally quite interested in CCC, and looked at it in a different way. He didn't look for higher or lower temperatures, but temperatures that were particularly uniform over the rings.

So this is a long story filled with lots of arguments and controversy based upon how one determines what is truly random or not. You see, you might see a feature and then ask yourself something like, "*I think there's a ring there, but maybe that's something caused by purely random fluctuations and has nothing to do with my theory at all?*"

So how could you determine that—how would you test it? The standard way people test it is by making "a fake sky"—a random sky—and see how many of these rings you find. And Vahe did this, and found that you see far more of these in the CMB than in a random sky.

But the key question is, *What do you mean by "a random sky"?* Many people who do these sorts of things mean a sky by which you take information taken from the observed sky—what we call the power spectrum, the general distribution of temperatures. And then you feed that into your model of "the random sky", essentially declaring it to be "random within that context".

But from our perspective this is a circular argument, because if these effects we're talking about are a major contribution to the CMB inhomogeneities, then it makes no sense to have those so integral to your definition of "randomness".

Well, I won't go into the whole argument, but I should just say that it's going on still, because even if you agree to put the power spectrum in to your model, there's still a question: *Do you put the actual observed power spectrum in, or do you put in the spectrum that you get from all the theory that you have?*

There's very precise agreement between the two for most of the curves, but when you look at the very fine details, it's not so clear: there are lots and lots of free parameters that you can put in to make it agree with the observed spectrum, and then you see more of these circles.

At any rate, as I said, this is still ongoing. It's the simulations that are the key issue for most people: have we done the simulations right? And there's a great deal of controversy about that: Vahe is convinced that he's done them correctly, while others are convinced that he hasn't.

So at some point, I thought about trying something else to move beyond this impasse with simulations. And I thought to myself, *What if we take the sky and twist it?*

In other words, we don't concern ourselves with simulations anymore and instead take the very precise CMB data from the WMAP and Planck satellite experiments and try to gauge how sensitive the data are to concentric circles, as opposed to ellipses.

We started by looking at the centers of at least four concentric rings. We found 56 of them. Then you twist them by an amount which makes the circle elliptical by about 1%—the ratio of major to minor axis is about $101/100$. That change reduces the number found in the CMB by a factor of about 4: you now see only about 10–20. If you increase the eccentricity to a much higher degree—say a 2:1 ratio of major axis to minor axis—you find that they completely vanish. So you see lots of concentric sets of 4 circles, but concentric sets of 2:1 ellipses you don't see at all.

It seems to me that that's fairly persuasive that there's something going on: whatever it is—even if it's not CCC—it's something that likes circles. It doesn't like even faintly distorted circles.

Questions for Discussion:

1. *To what extent is it possible to distinguish between "truly random" and "random within a certain context" when fundamental aspects of that very context are embedded in our complex theoretical models?*

2. *How might one envision building upon Roger's "twisting technique" to help confirm the claim that the CMB is particularly sensitive to circular patterns?*

X. Battling Onwards

Challenges of an irrepressibly independent thinker

HB: Let me move now to the position of a sceptical theorist. They might say something like, *"Well, you can always find whatever you want in the CMB data if you look hard enough and if you use the right sort of filter: you can pull out anything—you can pull out my grandmother if you're sufficiently diligent."* But, let's look at the theory itself and address the theory on its own grounds.

What's the biggest amount of pushback that you're getting in terms of the concepts and the ideas? What do people feel the most amount of discomfort with what you're saying?

RP: That's a good question. I've certainly had feedback on the fade out of mass. They say, *"Well, that's pretty artificial, isn't it?"* OK, I accept that: it's made to fit the theory. There's no evidence against it, but it is a fair point to say that it's artificial to introduce this, but the scheme seems to demand it.

There's a lot of scepticism about the observations. That's where the main doubts are, not so much the theory. They say, *"Well, you probably haven't done your analysis right."* There's a lot of that.

Then there's the fact that many people don't like to believe in information loss in black holes, which would be a big reason for doubting it.

Personally, I think that this issue is already there in standard cosmology: there is this remote future state, which, if you squash it back down through conformal mapping—which you can just do for fun—you get something very much like the Big Bang with the gravitational degrees of freedom not activated.

So there's a problem there. Black holes are where the entropy resides, and if it's come out in their evaporation, where has it gone in this picture? There's a big puzzle with this issue, quite apart from CCC, and I think you need to readdress the question of information loss in black holes to resolve it. But many people are uncomfortable with readdressing this: to say that it needs to be readdressed is currently the minority view.

Otherwise, I think a lot of it is inertia. You pick out any modern book on cosmology, popular or technical, and it will discuss inflation. You see that picture that everyone puts up of the funnel of the universe: inflation is right there at the beginning, the flat end. It's in everybody's imagination who takes cosmology seriously.

Many people don't like it. You don't have to be *me* to dislike the picture of inflation. It seems awfully artificial: it's highly contrived. You have to make the potential function do what you want—there's no theoretical reason for it to do that. And it leads to all these other problems people have with eternal inflation, which carries many more difficulties with it.

HB: Looking beyond CCC in particular, has there been more of a general appreciation of the entropy concerns of the early universe that have long been bothering you? Are people beginning to appreciate that this is a significant problem, a problem that inflation—whether we agree with it or not—doesn't actually address, and that significantly more work needs to be done on the theoretical side of things?

RP: I haven't seen a huge amount. The trouble is that I don't get the proper objective view. I go lots of places and give talks, and at the end people clap and say things like, "*I like your pictures,*" or "*I didn't understand the whole thing, but I hope you come back and give another talk sometime.*"

And I wonder to myself if they're not saying to each other, "*Has he gone off his nut, this fellow? He used to be a good scientist, and now he's talking about all these crazy ideas.*" I have a sort of feeling that a little of that's going on, you see. But I don't know, because people don't tell me directly. I have to rely on spies.

HB: But throughout your career you've often gone against the grain in many different ways.

RP: Yes. Perhaps they're just saying, *"He's just doing it again"*. I don't know. I need better spies.

HB: Well, thanks very much for giving me so much of your time, Roger. I have one final question before we finish: Are you sufficiently masochistic to have started another one of these large popular tomes?

RP: Well, yes. I started a book over a decade ago at the behest of Princeton University Press. They said to me, *"We'd like you to give three talks at Princeton and we'll make a book out of it afterwards"*, and they asked me for a good title. In a moment of idiocy, I thought of the title, *Fashion, Faith and Fantasy in the New Physics of the Universe*.

HB: It's wonderful alliteration.

RP: Yes, that's probably what made me stick to it, and what they liked about it. But the idiocy of it was that all this started off by lectures at Princeton.

What was the example of "fashion" that I had in mind? Well, string theory, of course. And who's at Princeton? Well, many of the great experts, including people like Ed Witten, with whom I had a very nice talk during my visit, although I don't think he was at that particular talk, and I'm sure he wasn't the slightest bit interested in what I had to say about it.

"Faith"? Well, that has to do with quantum mechanics at all levels: the complete faith people have in quantum mechanics. OK, it works wonderfully on the level that we use it, but if you're going to apply it to the universe as a whole, or even macroscopic bodies, there's an issue there.

And the "fantasy" part? Well, basically inflationary cosmology. So here I was at Princeton: an idiot going into this lion's den.

HB: I prefer to use the word "courageous".

RP: It's more "idiotic", I think. I keep imagining that if I speak about these things in ways which seem to me to hold together logically, then everyone will simply say, "*Oh yes, I see now. I was wrong all the time with my life's work.*" That's my idiocy.

HB: Well, I certainly beg to differ. And for my part, I've had a wonderful time, as expected, having the chance to get together and chat with you about CCC and other things. Thank you very much for letting me into your home and giving me the opportunity to talk with you in front of the cameras.

RP: It's been a great pleasure for me too.

Questions for Discussion:

1. Do you think that today's theoretical physicists are too conservative? Not conservative enough? Do different academic and geographical environments have different degrees of conservatism?

2. Are you surprised to learn that even a scientist as renowned and admired as Roger Penrose feels that he can't get an honest response to his ideas without "relying on spies"?

3. How, exactly, do you think that a colleague who disagrees with Roger would respond to his wide range of criticisms and scientific ideas expressed in this book?

Continuing the Conversation

Readers who enjoyed this discussion are strongly encouraged to read Roger's book, *Cycles of Time*, which goes into considerable additional detail about many of the issues discussed in this conversation.

Additional Ideas Roadshow conversations not offered in this collection that the reader might enjoy include *The Problems of Physics, Reconsidered* with University of Illinois Nobel Laureate **Tony Leggett** and *Cryptoreality* with University of Oxford and NUS quantum information theorist **Artur Ekert**.

Physics, Continued

Ideas Roadshow offers several additional conversations with leading physicists available both individually and as 5-conversation collections, such as *Conversations About Physics, Volume 1* and *Volume 2*. A full listing of all titles can be found at:

www.ideas-on-film.com/ideasroadshow.

Printed in Great Britain
by Amazon